Zero Carbon Britain: Rethinking the Future

Zero Carbon Britain:
Rethinking the Future

www.zerocarbonbritain.org

Primary authors: Paul Allen, Laura Blake, Peter Harper, Alice Hooker-Stroud, Philip James and Tobi Kellner.

Contributing authors: Vijay Bhopal, Guppi Bola, Isabel Bottoms, Julie Bromilow (CAT Education), Louisa Casson, Richard Hebditch, Ling Li, Nuria Mera-Chouza, Lucy Neal, Darcy Pimblett, Sophie Wynne-Jones, Liz Zeidler and Mike Zeidler.

Machynlleth, Powys
SY20 9AZ, UK
Tel. 01654 705950 • Fax. 01654 702782
info@cat.org.uk • www.cat.org.uk

ISBN: 978-1-902175-69-0
1 2 3 4 5 6 7 8 9

Editors: Alice Hooker-Stroud, Hele Oakley and Allan Shepherd.
Editorial support: Sarah Everitt, Megan Jones and Rebecca Sullivan.
Typesetting, and layout: Graham Preston (grahamjpreston@gmail.com).
Illustrations and cover design: John Urry.
Illustration support: Sarah Everitt and Megan Jones.
Photography: As credited, except those photographs which appear on chapter heading pages, which are © CAT (Executive summary, Chapters 1 and 5, and End notes) and Joanna Wright (Chapters 2, 3 and 4).
Publishing: Allan Shepherd, on behalf of the Centre for Alternative Technology.
Proofreading: Rebecca Sullivan.

Published by CAT Publications, CAT Charity Ltd.
Registered charity no. 265239.

Printed in the UK by Cambrian Printers, Aberystwyth, 01970 627111.

Amadeus is made from 100% virgin fibre which is sourced from sustainable forestry guaranteed for peace of mind via an international chain of custody certification.

Dedication to Richard St George

On behalf of the Zero Carbon Britain project I would like to dedicate this new report to the memory of our good friend and colleague, Richard St George. One of CAT's early pioneers, Richard was directly involved with the communication of CAT's original 1977 report, *An Alternative Energy Strategy for the UK*. He remained a good friend of CAT and a keen collaborator in his subsequent role as director of the Schumacher Society.

Richard was also highly influential in catalysing the current Zero Carbon Britain project in 2007. Richard's enthusiasm for 'getting things off the ground' meant he would regularly pull together meetings of the 'Schumacher Circle' of organisations, which included CAT, the new economics foundation, Practical Action and the Soil Association.

It was at one of these meeting that we recognised an urgent need for a 'Marshall Plan'– to show what positive green futures could be like if we made all the right choices. It was this meeting, coupled with Richard's 'can do' enthusiasm, that gave me the inspiration and confidence to re-visit the CAT 1977 report and establish a new research team that resulted in our first Zero Carbon Britain report in 2007. Richard also played a key role in getting the Zero Carbon Britain message out. In his role as organiser of the UK Schumacher Lectures, he facilitated the running of a conference centred on the findings of the second report (*ZeroCarbonBritain2030: A New Energy Strategy*) in 2010, enabling us to reach a wide range of radical green thinkers.

We are proud to dedicate this new work to his memory.

Photo kindly provided by the family of Richard St George

Paul Allen
Zero Carbon Britain Project Co-ordinator

Acknowledgements

Many wonderful people helped and contributed in a great many ways to this project – too many to name individually here.

Here we list some who generously gave time and expertise to help ensure an up to date and high quality piece of research. Thank you to the many individuals who helped directly with our research by attending our seminars and conferences, answering questions and reviewing our work, providing articles and data, and generally pointing us in the right direction:

Prof Kevin Anderson, Eric Audsley, Clifton Bain, Dr Tom Barker, Dr John Barton, Dr Jessica Bellarby, Mike Berners-Lee, Dr Brenda Boardman, Dr Alice Bows, Prof Godfrey Boyle, Dr Arthur Butler, Shaun Chamberlain, Kevin Coleman, Jane Davidson, Sue Dibb, Prof Dave Elliott, Dr Chris Evans, Dr Kerrie Farrar, Dr Tina Fawcett, Dr David Finney, Sue Fowler, Dr Tara Garnett, Jim Hammond, Sara Hartwell, Prof Tim Lang, Dr Robert Matthews, Prof Erik Millstone, Daniel Quiggin, Tim Randle, Giles Ranyl Rhydwen, Piers Sadler, Dr Simon J. Shakley, Prof Pete Smith, Dr Saran P. Sohi, Dr Mark Stringer, Dr Murray Thomson, Dr Ruth Wood, Dr Fred Worrall, Dr Adrian Williams, Duncan Williamson, and Prof John Wisemann.

Thank you also to our dedicated production team for your patience, support, understanding and skill – Allan Shepherd, John Urry, Graham Preston, Hele Oakley and Rebecca Sullivan.

Responsibility for any errors, omissions or mistakes, however, lies solely with the Zero Carbon Britain project as part of the Centre for Alternative Technology.

We would also like to say a huge thank you to our friends and families for their support and encouragement, and to our colleagues here at the Centre for Alternative Technology (CAT) – not forgetting all those who volunteer – for their enthusiasm for the Zero Carbon Britain project and our research, for their care and compassion to us as individuals, and their heartfelt dedication to the ethos of CAT and the work we all do here.

Last, but not least, we would like to thank those individuals and organisations who donated generously, enabling this project to go ahead. They are, in no particular order:

CAT members and supporters, the William A. Cadbury Charitable Trust, the Marmot Charitable Trust, the Polden Puckham Charitable Foundation, the W. F. Southall Trust, Jam Today and, of course, Richard St George, to whom this report is dedicated.

Foreword

"As the need to reduce our carbon emissions becomes ever more urgent, this well researched and timely report makes a key contribution to the debate.

The challenge is to resolve the growing disconnect between what scientists tell us is needed and what policymakers tell us is possible. It is worrying that as the scale of the problem increases, public concern seemingly falls, and the disconnect grows ever larger.

With this report, the Centre for Alternative Technology once again places itself at the forefront of informed debate by showing how practical action can bridge this gap.

By setting out what a low carbon world would look like it shows that the solutions to our problems do exist and all that is needed is the political will to implement them. We must create grassroots pressure on politicians to recognise the scale of the problem and to rise to the challenge. Not only is this essential for a sustainable future but vital for our sense of wellbeing.

This report is essential reading for politicians, policymakers and anyone interested in developing effective solutions to our climate problems. As Chair of the Environmental Audit Select Committee and of the All Party Climate Change Group, I will do all I can to raise awareness amongst my parliamentary colleagues and I encourage everyone else to make whatever contribution they can to ensure we pass on a sustainable world to future generations."

Joan Walley MP
Chair of the Environmental Audit Select Committee
Chair of the All Party Parliamentary Climate Change Group

"The last report published by CAT in 2010 – *ZeroCarbonBritain2030: A New Energy Strategy* – was well received and contained valuable arguments and information. However, despite the challenges outlined in that and many other reports on climate, industrial countries like our own have not taken on board the urgent need for action to turn the continuing growth in carbon emissions into substantial year-on-year reductions.

There are three areas where reductions in emissions must be particularly sought. The first is in agriculture and land use, related to our food production and diets. The second is improved energy efficiency in our existing building stock and infrastructure. The third is a much more rapid development of new renewable sources of energy. Effort in these areas will also bring many co-benefits, for instance improvements in health and employment opportunities.

I strongly recommend the new Zero Carbon Britain report and trust that it will lead to serious and effective action."

Sir John Houghton
Former Co-chair of the Intergovernmental Panel on Climate Change (IPCC)
Former Director General of the UK Met Office and founder of the Hadley Centre

About the Zero Carbon Britain team

Paul Allen – Project Co-ordinator. Holding an Honours degree in Electronic and Electrical Engineering, Paul joined the Centre for Alternative Technology (CAT) in 1988. He assisted the development and production of a wide range of renewable energy systems and helped develop CAT's spin-off engineering company, Dulas Ltd. He is currently CAT External Relations Officer and ZCB Project Co-ordinator. Paul is also Member of the Wales Science Advisory Council (2010), board member of the International Forum for Sustainable Energy (2008) and a Climate Change Commissioner for Wales (2007).

Under Paul's co-ordination of the ZCB project, the following research team was assembled in the summer of 2012 at the Centre for Alternative Technology (CAT). This team undertook the majority of the research that has gone into the creation of the scenario for this report.

Alice Hooker-Stroud – Research Co-ordinator. Alice has Master's degrees in Physics and in Earth Systems Science. She has contributed to work modelling fossil fuel CO_2 emissions; helped set up a research group in Barcelona, Spain; and was the primary author of a set of climate science factsheets designed for campaign groups. She believes that evidence-led policy supported by society is required to make the transition to a sustainable world. She currently enjoys living in – and her ability to correctly pronounce – Machynlleth, in Mid Wales.

Tobi Kellner – Energy Modeller. Tobi heads CAT's renewable energy consultancy service and lectures on energy related courses. He has a background in computer science and is now especially interested in how we can integrate fluctuating renewable energy sources into a stable energy system that 'keeps the lights on' when the wind isn't blowing. When not modelling energy flows he forages for stinging nettles and tries, unsuccessfully, to grow edible mushrooms.

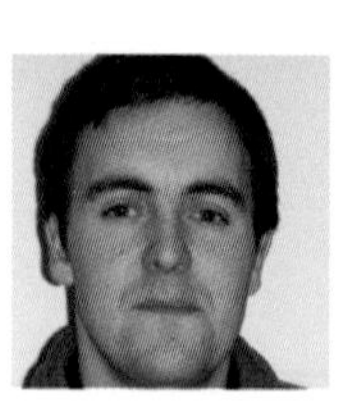

Philip James – Energy Systems Researcher. Having completed a doctorate studying strategies for low and zero carbon homes, Philip was keen to get involved in research looking at the energy system as a whole. He is interested in the ways we can reduce our energy demand and make it more flexible, and is working to understand how we can match demand with a decarbonised energy supply. As part of ZCB he has also really enjoyed researching how our future energy and land use systems might interact.

Peter Harper – Food and Diets Lead Researcher. Peter Harper is semi-retired from CAT, having worked here for 30 years. He was a pioneer of the alternative technology movement – coining the term in 1972 – and is still very much a part of it. While at CAT he combined writing, teaching and research in many fields, including horticulture, waste treatment, carbon accounting, sustainable development and environmental history. Since 2007 he has worked with the Zero Carbon Britain project, focusing on the large-scale infrastructural changes required for rapid decarbonisation.

Laura Blake – Food and Diets Researcher. Laura has an undergraduate degree in Nutrition and an MSc in Food Security. During her studies she covered topics such as public health, agriculture, climate change and food governance. She researched obesity and developing countries (the nutrition transition); and the interactions among global human population, livestock production systems and sustainability. She believes that food issues should be tackled holistically – addressing health requirements, social justice issues and environmental sustainability together.

Joanna Wright – Artist in Residence. ZCB won funding from the Arts Council of Wales this year for Joanna to join the team as artist in residence. She is a Welsh artist working with documentary film, oral history recording, archives and photography. She lives and works in Bangor, north-west Wales.

Finally, a number of interns, trainees and volunteers all worked with the ZCB team taking responsibility for a wide variety of roles: editorial work, research and writing, media and communications, and helping to feed ZCB into other parts of CAT's work. They are:

Nuria Mera-Chouza – Leonardo Da Vinci programme trainee (Energy Systems Assistant Researcher). Nuria studied Chemical Engineering at the University of Cádiz, Spain. She has studied courses on energy use, management and efficiency. Whilst with the ZCB team, she researched the Fischer-Tropsch and Sabatier chemical processes.

Jan Lahobý – Intern (research on European decarbonisation scenarios). Jan studied political, media and environmental sciences at the Faculty of Social Studies at Masaryk University in Brno, Czech Republic. He has been working in the Ecological Institute Veronica as the Climate Protection Co-ordinator since the beginning of 2011.

Richard Delahay – Volunteer (ZCB-CAT Media and Marketing Liaison). Richard now works at CAT's visitor centre and volunteered for the ZCB project to communicate its positive message for sustainable communities.

Megan Jones – Volunteer (ZCB-CAT Media and Marketing Liaison). Megan has been doing conservation work, environmental research and has been involved in activism for several years. She has a BA in English from Reed College in Portland, Oregon.

Sarah Everitt – Volunteer (ZCB-CAT Education Liaison). Sarah is an Environmental Science graduate passionate about working towards a sustainable society. She has enjoyed working on a range of voluntary environmental projects, now including a long-term placement here at CAT where she supported the ZCB team in producing this report.

Ling Li – Volunteer (ZCB Policy Researcher). Ling has worked for five years managing and developing emissions reductions projects and researching carbon markets and related policy. She received her Master's in Environmental Management from Duke University in the USA.

Contents

Executive summary 1

Chapter 1 Introduction 5
- 1.1 History of the Centre for Alternative Technology 6
- 1.2 History of the Zero Carbon Britain project 7
- 1.3 Why a third ZCB report? 9
- 1.4 What ZCB is and is not 10

Chapter 2 Context 11
- 2.1 The global situation 12
  - 2.1.1 So you think this is normal? 12
  - 2.1.2 Climate change 13
  - 2.1.3 Planetary boundaries 17
  - 2.1.4 Future generations 19
- 2.2 The situation in the long-industrialised west 20
  - 2.2.1 Energy supplies 21
  - 2.2.2 The economic crisis 22
  - 2.2.3 Wellbeing 22
- 2.3 What does this mean for the UK? 24
  - 2.3.1 Our carbon budget 25
  - 2.3.2 The physics-politics gap 27

Chapter 3 Our scenario: Rethinking the Future 29
- 3.1 About our scenario 30
  - 3.1.1 Aims 30
  - 3.1.2 Rules 31
  - 3.1.3 Assumptions 34
- 3.2 Measuring up today 35
- 3.3 Power Down 38
  - 3.3.1 Buildings and industry 40
  - 3.3.2 Transport 47
- 3.4 Power Up 54
  - 3.4.1 Renewable energy supply 56
  - 3.4.2 Balancing supply and demand 63
  - 3.4.3 Transport and industrial fuels 70
- 3.5 Non-energy emissions 73
  - 3.5.1 Industry, businesses and households 74
  - 3.5.2 Waste 76
- 3.6 Land use 81
  - 3.6.1 Agriculture, food and diets 83
  - 3.6.2 Growing energy and fuel 94
  - 3.6.3 Capturing carbon 98
- 3.7 Measuring up 2030 108
- 3.8 How we get there 111
  - 3.8.1 ZCB and the UK's carbon budget 111
  - 3.8.2 Zero carbon policy 114
  - 3.8.3 Economic transition 118
- 3.9 Benefits beyond carbon 120
  - 3.9.1 Adaptation 121
  - 3.9.2 Planetary boundaries 124
  - 3.9.3 Employment 126
  - 3.9.4 Wellbeing – measuring what matters 128
- 3.10 Other scenarios 130
  - 3.10.1 Scenario variations using ZCB rules 131
  - 3.10.2 Breaking the ZCB rules 133
  - 3.10.3 Carbon omissions 134

Chapter 4 Using ZCB 137
- 4.1 Changing how we think about human beings and energy 138
- 4.2 Taking action in our homes, communities and places of work 140
- 4.3 Influencing policy 143
- 4.4 Zero carbon education 144
- 4.5 Developing a zero carbon project 145
- 4.6 Reclaim the future: engaging with arts and creative practice 147

Chapter 5 ZCB and ... 149
- 5.1 ZCB and drivers 152
- 5.2 ZCB and community energy 154
- 5.3 ZCB and farmers 156
- 5.4 ZCB and health 158
- 5.5 ZCB and young people 160
- 5.6 ZCB and happiness 162
- 5.7 ZCB and Zero Carbon Egypt 164
- 5.8 ZCB and creative practice 166

End notes, glossary and references 169
- Find out more 171
- Notes 171
- Glossary 173
- References 179
- Index 198

Executive summary

Zero Carbon Britain: *Rethinking the Future* describes a scenario in which the UK has risen to the challenges of the 21st century. It is 2030. We have acknowledged our historical responsibility as a long-industrialised nation and made our contribution to addressing climate change by reducing UK greenhouse gas emissions rapidly to net zero.

Our research shows that we can do this without relying on promises of future technology, but by using what exists now. By making changes to our buildings, transport systems and behaviour, and by investing in a variety of renewable energy generation technologies suited to the UK (without a nuclear component), we can provide a reliable zero carbon energy supply without negatively impacting on quality of life. Smart demand management, plus the intelligent use of surplus electricity in combination with biomass to create carbon neutral synthetic gas and liquid fuels, mean that we can meet our entire energy demand without imports, and also provide for some transport and industrial processes that cannot run on electricity.

In our scenario the biomass we require is provided by growing second generation energy crops on UK land. All our cropland is still used for food production, and we produce the vast majority of the food required to provide for the UK population on home soil. Changing what we eat (mainly a significant reduction in meat and dairy products, coupled with increases in various other food sources) means we eat a more healthy and balanced diet than we do today while our agricultural system emits fewer greenhouse gases and uses less land both at home and abroad, thus decreasing the environmental impact of our food production globally.

We balance out some greenhouse gas emissions that cannot currently be eliminated from non-energy processes (industry, waste and agriculture) by using safe, sustainable and reliable methods of capturing carbon. By restoring important habitats such as peatland, and by substantially expanding forested areas, we not only capture carbon but also provide wood products for buildings and infrastructure, rich environments for biodiversity and more natural spaces for all of us to enjoy.

An initial analysis shows that, in this future, our actions have also helped us adapt to expected changes in climate while increasing our resilience to unexpected changes; improved upon a number of other significant environmental problems aside from climate change; created over a million jobs; and have had a positive impact on our economy and on the health and wellbeing of individuals and society.

The key difference between this future scenario and that for which we are currently heading is that we have responded with the urgency demanded by current climate change science, taking a physically realistic perspective rather than adhering to what might be politically or socially palatable today. It is unethical to treat fundamental needs in the future, and the needs of others in the global community, as equivalent to our lifestyle preferences in the West today.

Current UK greenhouse gas emissions targets, though ambitious in comparison to our international contemporaries, do not offer substantial enough reductions to provide a good chance of avoiding what is now considered extremely dangerous climate change. Neither do they adhere to what might be termed the UK's 'fair share' of the global carbon budget. The most recent climate science now demands a much greater sense of urgency than the current mainstream view.

Zero Carbon Britain: Rethinking the Future explores how we can achieve what is necessary. Building upon the groundwork laid by the Zero Carbon Britain project over the last six years, we incorporate the latest developments in science and technology, plus more detailed research in two main areas: balancing highly variable energy supply and demand; and the nutritional implications of a low carbon diet (see box). We highlight the need for further research on adaptation, economic transition and policy that would achieve sufficient greenhouse gas emissions reductions quickly and equitably. From a broader viewpoint, we also highlight the need to incorporate greenhouse gas emissions associated with our 'historical responsibility' as a long-industrialised nation, and with the goods and services that we import ('carbon omissions'), into international policy negotiations.

Closing the gap between current 'politics as usual' and what is physically necessary to address climate change will require cross-sector collaboration and public engagement, framed by robust international agreements to foster high-level all-party political commitment. *Zero Carbon Britain: Rethinking the Future* provides a positive and technically feasible future scenario that aims to stimulate debate and catalyse action across all parts of society. Through this project, the Centre for Alternative Technology (CAT) hopes to inform, inspire and enable contemporary society to embrace the changes required to rethink the future.

Practical advice on being part of the transition to a zero carbon Britain and further discussion papers written by a variety of individuals and organisations are featured at the end of this report. We invite you to explore your own reflections on life in a zero carbon Britain and to get involved in creating and working towards a positive future.

Overview of this research phase

Two new pieces of research underpin the development of this scenario:

1) Hourly modelling of the UK energy system in our scenario using ten years of weather data to simulate our renewable electricity supply (wind speed, sunlight, etc.); and the demand for electricity during, for example, periods of cold and warm weather (temperature).
   Even with a significantly reduced energy demand and a broad mix of renewable electricity generation technologies, supply and demand do not change in unison – there are times when our energy systems produce a surplus, and others when they fall short of demand. Our hourly modelling research shows that this imbalance can be managed with a combination of demand management techniques, some short-term energy storage, and the provision of a small amount of back up generation.
   One important outcome of this research is the need for dispatchable energy over baseload power. Constant power output (such as that from nuclear power plants) is not helpful in balancing a variable energy supply – it simply leads to further overproduction of energy at times when renewable systems can meet demand. We require instead power from generators that can very flexibly increase or decrease output, or even switch off completely, depending on whether or not renewable sources are catering for demand. Present gas infrastructure, including storage facilities and gas power stations that can quickly ramp up output, provide the best solution for this, and can be made completely carbon neutral – using synthetic gas created with surplus electricity from renewables and UK-grown biomass.
2) Modelling of low and minimal carbon diets. Dietary analysis based on nutritional profiling, food group balance and government dietary recommendations enables us to provide a healthy average diet for the UK while monitoring the implications of various dietary choices on greenhouse gas emissions and land use requirements.
   Today's average UK diet contributes not only to multiple environmental issues at home and abroad, but also to an increasingly unhealthy population which suffers from a multitude of diet-related diseases. Currently, we in the UK overeat and lack balance in our diet.
   The principle outcome of our research is that, in general, a healthier diet is also lower in greenhouse gas emissions, and demands less of our land. This win-win-win situation takes a pivotal role in our scenario, not only providing multiple impetuses for dietary changes, but importantly releasing land for other uses – providing biomass for our energy system, and safe and proven carbon capture to balance our remaining emissions.

Chapter 1

Introduction

1.1 History of the Centre for Alternative Technology (CAT)

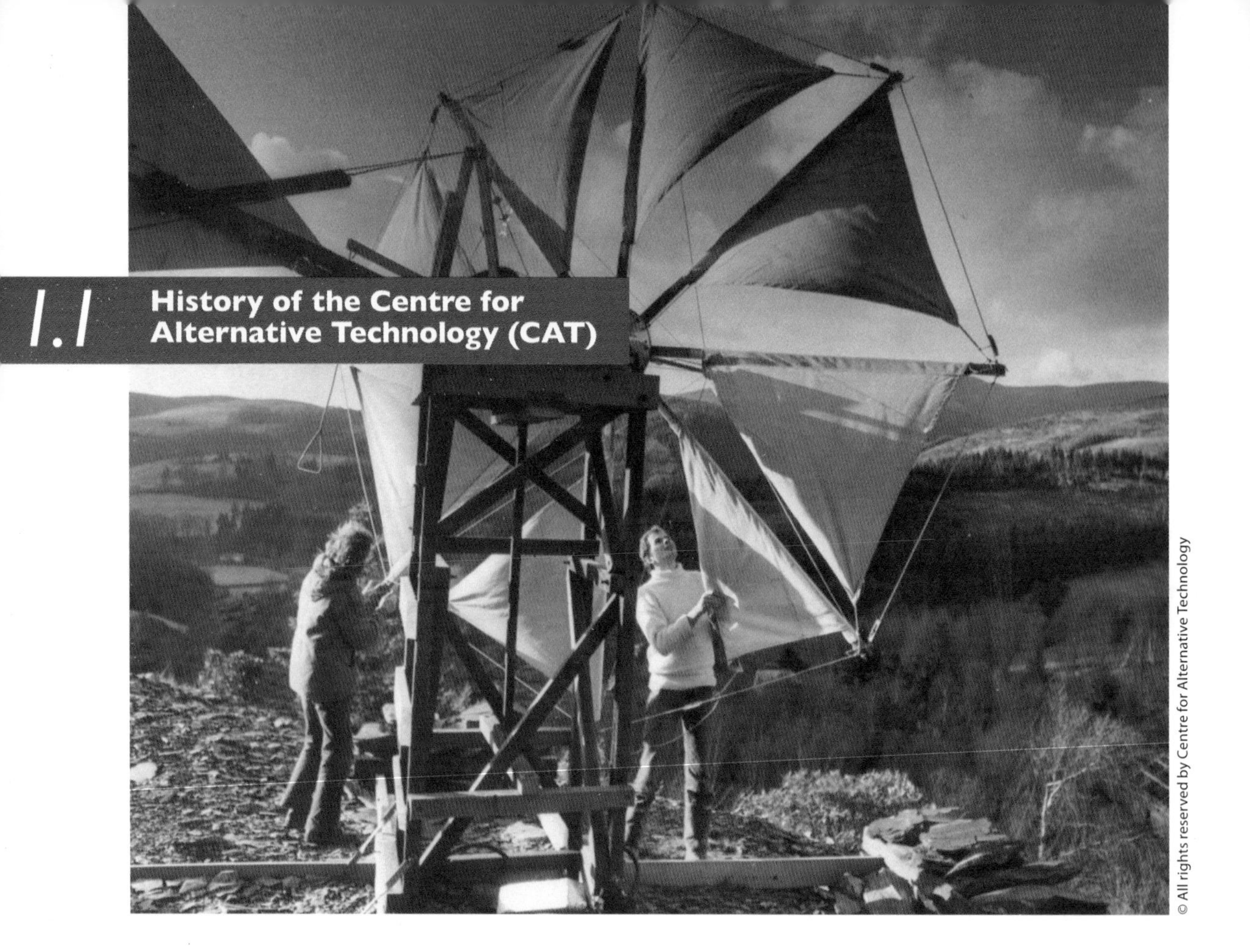

"In the early 1970s I took a sabbatical and went to America. I talked to senior business and professional people and came to the conclusion that a lot of people realised there was a major problem, but were locked into what they were doing. I came back thinking what was needed was a project to show the nature of the problem and to indicate ways of going forward."
Gerard Morgan Grenville – CAT Founder

Forty years ago, catalysed by Gerard Morgan Grenville's vision, a small group of young visionaries adopted a long-derelict slate quarry in the village of Pantperthog, near Machynlleth in Mid Wales.

At the time, an important shift in the relationship between human beings and technology was happening. Until then, developments in technology were seen to bring progress and an ever-improving standard of living, and had been largely unquestioned as a result. However, as the industrial world began to collide with the limits of the planet's ecosystems, serious questions arose about the limits to material growth, damage to natural systems and the eventual depletion of resources.

This rethinking of the direction of science and technology gave rise to a key conference at which Peter Harper coined the phrase 'alternative technology' to describe a new role for technology, focusing on benefits to humans and nature as well as to economies. Alternative technology wasn't just about solar and wind power, but rather a shift in the philosophy of how a technology is applied and to what ends. Gerard took this concept as the basis for developing the Centre for Alternative Technology (CAT).

Society was just emerging from the swinging sixties, and few people were watching the problems, let alone looking for the solutions. This original community set out to test and develop, by a positive living example, new technologies that could provide practical solutions to problems now worrying the world's ecologists, economists and energy analysts. These early pioneers began trying out a wide range of low-impact, self-sufficient or self-reliant technologies, such as growing, cooking, nutrition, alternative medicine, clothing, buildings, smallholdings, transport, foundry skills, wildlife management and co-operative decision-making. This hands-on research would not only further the all round 'living lightly' message, it would feed, house, clothe, power and manage the community, independent of the mainstream.

Right from the outset, however, CAT recognised that building a genuinely sustainable future would need thousands of skilled professionals with a deep understanding of environmental technologies and practices.

Today, CAT offers residential courses, taught by experts with many years of practical experience, based in a 'living laboratory' with a new state of the art teaching facility. CAT's Graduate School of the Environment (GSE) offers research, training and hands-on skills up to postgraduate level, with core topics including low carbon building techniques, grid-linked and stand-alone renewable energy, solar water heating, ecological building, eco-renovation, sewage treatment, water supply, organic food production, composting, architecture, adaptation and solid waste disposal – each exploring the complex interaction between land use planning, food production, energy, buildings, transport, waste management and all aspects of human society.

.2 History of the Zero Carbon Britain project

A key part of Gerard Morgan Grenville's original vision was for the CAT project team to assemble the findings of its research by the end of the first five years. These findings were to describe what the emerging alternative technologies could realistically offer. This was completed in 1977 with a report for the UK government.

The initial vision – *An Alternative Energy Strategy for the UK* (1977)

Experts from CAT initiated a process of collaboration, embracing leading thinkers from a number of key universities and industries. This led to the production of the very first *Alternative Energy Strategy for the UK*. Sixteen copies were bundled up and delivered to Tony Benn's Ministry of Energy. Not surprisingly, the reception from the energy mainstream varied from scorn to outright hostility. The strategy was poles apart from that of the official energy strategy of the time.

Back in the early 1970s, the majority of mainstream energy planners expected UK demand to grow year-on-year as it had been doing since the end of the Second World War. This continued expansion in energy consumption was to be fuelled by the as yet untapped North Sea oil reserves and the promise of nuclear power, which was going to be "so cheap it wouldn't be worth metering". Renewable energy played a very small part in the national energy mix. Wind power and hydropower energy systems were associated with remote 'pre-national grid systems' used by remote rural villages in the 1920s and 1930s.

The national grid, managed by the Central Electricity Generating Board, was not interested in buying power from any suppliers with a capacity below 10 megawatts.

The oil price shocks of 1973 and 1979 gave a jolt to the mainstream, but they were portrayed as short-term political problems rather than precursors of overall resource depletion. They did, however, motivate the alternative movement – looking wider and further ahead than the mainstream. CAT's innovative report showed for the very first time that an alternative approach could head-off resource depletion by reducing energy demand whilst radically increasing generation from renewable sources.

The first ZCB report (2007)

Throughout the last decades of the 20th century, evidence of a different threat had been building – that of climate change. By the start of the 21st century, the importance of taking action to deal with this new challenge had grown ever more urgent. However, efforts were still focused on communicating the problem. Research at that time showed that 60% of articles about climate change in UK national newspapers were negative and failed to mention possible solutions; only a quarter mentioned what could be done or was already being done.

At that time, the UK official target (60% reduction of greenhouse gas (GHG) emissions by 2050) fell far short of what science was demanding. Furthermore, no other published work put forward decarbonisation scenarios that explored a fast enough transition from fossil fuel use to meet the challenge. Although a number of groups had developed scenarios around a decarbonised electricity grid for the UK, they did not cover GHG emissions from non-electrical energy demand – which is by far the largest part of current UK energy demand.

The challenges of climate change, fossil fuel depletion and global inequality had become increasingly familiar individually, but experts worked in isolation and their solutions were rarely considered in unison. CAT sought to develop a scenario that could integrate solutions to all of these challenges.

In collaboration with our MSc tutors and students, CAT began a series of consultations with the aim of creating a rapid decarbonisation scenario – one that included all of the UK's energy demands, including transport and heating as well as electricity. It would integrate a wide spectrum of cutting-edge research and practical hands-on experience to help increase the level and quality of the national energy policy debate, just as the original publication had done in 1977. It would propose solutions that tackled the challenges of climate, energy and inequality together.

The first report – *zerocarbonbritain: An Alternative Energy Strategy* – was a first pass over a new and unfamiliar energy landscape. It offered a scenario outlining a pathway to zero emissions in two decades utilising only proven technologies. We demonstrated a dual process of 'Powering Down' energy demand, and 'Powering Up' renewable energy supplies. Through a policy framework and technology scenario, the report provided the UK policymaking community with a vision for a truly sustainable energy future.

Within eight weeks of its parliamentary launch in Westminster, the UK Liberal Democrats launched their own energy policy: 'Zero Carbon Britain – Taking a Global Lead'. Lembit Opik MP, then Liberal Democrat Shadow Secretary of State for Business and Enterprise, acknowledged that: *"These proposals were largely inspired by the Welsh based Centre for Alternative Technology and I would like to thank them for the ground-breaking work."*

The second ZCB report (2010)

The economic meltdown began almost immediately after the launch of the 2007 report. The ensuing chaos made it much harder to make the case for concerted action on climate change. It became clear that the scenario must also highlight economic and employment benefits.

There was also recognition that to be truly zero carbon, all UK GHG emissions must be addressed – including those unrelated to energy. This proved a much harder challenge, with some emissions impossible to reduce to zero. The second report integrated new research in land use, resulting in a change in the role of land in the UK. Land in the scenario became of crucial importance, providing food, energy, fuel and, in particular, carbon capture – integral to making the scenario reach net zero carbon emissions.

What developed was a more robust framework that integrated detailed knowledge and cutting-edge research in transport, food, energy, buildings and land use and also demonstrated the potential economic and employment benefits.

"Zero Carbon Britain has allowed us to open up crucial conversations with government, Parliament, and the business world about Britain's energy future."
Catherine Martin, Co-ordinator, All Party Parliamentary Climate Change Group (2010)

In June 2010, the findings of this research were launched in the report *ZeroCarbonBritain2030: A New Energy Strategy,* at the All Party Parliamentary Climate Change Group by former IPCC (Intergovernmental Panel on Climate Change) co-chair and Nobel Peace Prize winner, Sir John Houghton.

Why a third ZCB report?

Since the 2010 report was published, the economic situation has worsened. International negotiations on climate change have stalled, and agreement between nations leading to deep emissions cuts on a global scale seems as distant as ever.

The UK finds itself in a catch-22 situation in which business, government and civil society are all looking to each other for leadership on climate. We as a society are perhaps not convinced of the need to act, as government action does not reflect the scale of the threat outlined by science. Government is hesitant to take bolder action out of concern that they lack the social mandate to do so. Business looks to government for certainty that policy will offer long-term stability for investment in decarbonisation and developing low carbon products, jobs and skills.

These interdependencies have prevented action at the scale and speed necessary to tackle climate change. Despite urgent new evidence, far too little progress has been made.

There is simply no historical precedent for the scale of the challenge we currently face. We lack the collective psychological and emotional tools required to understand or to react. In our experience, human beings (including us here at CAT) are naturally drawn to the immediate concerns of 'what we need to do now' and it takes a conscious effort to shift to exploring 'where we need to be' in the future.

This third report – *Rethinking the Future* – outlines 'where Britain could be if we rose to the challenge'. Addressing the physical realities of what science demands requires us to look beyond 'politics as usual' and what is currently socially palatable; it is clear that rising to the challenge will require major changes from within our society and democracy as well as from our technology. We offer a robust scenario that integrates cutting-edge knowledge and experience from a wide range of disciplines. It is a scenario showing that Britain can be a zero carbon society and gain benefits to health and wellbeing along with it. We address concerns about 'keeping the lights on' under a variable renewable energy supply, and 'feeding ourselves properly' on a low carbon diet; we look outside the realm of decarbonisation and discuss wider implications for our society and our environment.

We can meet the scale and speed of decarbonisation required with positive effects on society, the environment and the economy. We can acknowledge our historical responsibility as a long industrialised nation, and perhaps begin to break the deadlock to help catalyse global action on climate change.

What ZCB is and is not

Through researching and communicating this new report, CAT aims to stimulate economic and political debate around rapid decarbonisation, engage the research community and get society thinking in a new way to help build consensus on action. We do not intend this new scenario to be seen as the 'only way to save the planet' or some new eco-scripture that has to be followed to the letter. Our intention is to open conversations that start with the physical realities of what scientific consensus demands, acknowledging the UK's historical responsibility as a long industrialised nation that has been emitting GHGs for over 150 years. We aim to integrate a wide range of current research to show a possible scenario for rising to that challenge.

Exploring what it would be like to live in a Britain where we have risen to these challenges offers the potential to dispel myths, break through misunderstandings and trigger further research and collaboration on many topics pertinent to preventing further climate change – from new technologies to social science. By getting people thinking differently about the future, we hope to catalyse urgent action across all sectors of society – if we can't picture a solution, we will surely stay stuck in the problem.

Chapter 2

Context

marfis75 / Creative Commons

2.1 The global situation

Globally, our situation has changed phenomenally over the last 150 years or so. The advent of fossil-fuelled energy has enabled huge developments in science and technology, bringing massive benefits for our health, education and development.

But it is arguable that these benefits are not universal. Not all people have had the privilege of such developments, and inequality persists between (and within) nations. The environment on which we all depend has been adversely affected. We have put our future at risk of the very serious, dangerous and real consequences of climate change, mass extinction and global economic collapse.

Over the next 150 years, we will see the natural environment reacting to our actions, and the societal and cultural implications of our choices. We are butting up against the environmental limits of our planet, expecting continual development and growth in a finite world.

2.1.1 So you think this is normal?

In order to move forward we must understand the psychology of our current collective addiction to fossil fuels and how we were so gladly driven into the habit. On any historical or even geographical comparison, the amount of energy we now use in the developed West is highly abnormal, yet it has become normalised. A whole generation has grown up assuming the lights will come on, that there will be petrol in the pumps and stocked shelves in the supermarket. The challenge we face is not only for our technology, but also for our culture. Rising to this challenge requires us to consider the current relationship between human beings and energy in its wider historical context.

The story of human beings and energy began over 400 million years ago with the formation of fossil fuels. For millions upon million of years, plant life on planet Earth soaked up the sun's energy for photosynthesis, creating the largest, most concentrated and most convenient energy store we are ever likely to know.

Until relatively recently, we had no idea this energy store was under our feet. Our access to energy was limited to an annual ration of sunlight that reached the Earth's surface – providing the energy for plants to grow, making the wind blow and driving the water cycle. Access to land was vital – providing us with food to eat and fuel to keep warm. Over centuries we became more inventive, taking advantage of the trade winds to sail ships, and of wind and water to power windmills and waterwheels. All of this, however, still relied on the sun's annual energy ration.

The discovery of fossil fuels towards the beginning of the 19th century changed everything. With a powerful mix of the right skills and accessible stores, Britain burst into action with coal extraction, leading the world towards ways of making faster and larger withdrawals from a seemingly limitless account of ancient solar energy – fossil fuels. For the first time in human history, we had access to energy independent of land or season. Major changes in agriculture, manufacturing and transportation spread across Britain, Europe, North America and eventually the world. Oil soon displaced coal as the largest source of energy, being both easier to access and more transportable.

By the 1900s, the world was awash with abundant, cheap fossil fuels. Industrial and manufacturing processes were developed with little regard for the amount of energy they consumed. Continued expansion of access to fossil fuel energy gave rise to ever-growing industries. Our economic systems were built on the assumption that growth is the norm, and that it would be both perpetual and unrestricted.

Fossil fuel production was highly profitable, so much of our infrastructure was designed, quite literally, to use as much fossil fuel as possible. But at no time was this 'designed dependence' on fossil fuels as marked as with the arrival of the motorcar. Car production was to be the engine of post-Second World War economies; tramways were scrapped, rail links removed and newly sprawling towns and suburbs were deliberately developed in such a way that the car became not just a convenience but an absolute necessity.

Although the practice of having more than we need in order to highlight social standing in society is as old as civilisation itself, fossil fuels allowed this elite habit to become a mass culture. Conspicuous consumerism now exerts an irresistible pressure, making society reluctant to question the access to the energy supplies that underpins it.

Almost without realising it, we now depend on fossil fuels in nearly every aspect of our lives, while around the world they are linked to progress and betterment.

2.1.2 Climate change

When we burn fossil fuels to heat our homes and drive our cars, or when we use chemical processes in industry, change how we use land and produce the food we eat, greenhouse gases (GHGs), such as carbon dioxide (CO_2), methane (CH_4) and nitrous oxide (NO_2), are emitted. The burning of fossil fuels contributes most to greenhouse gas emissions (Baumert et al., 2005)

Even though plants and oceans absorb much of the CO_2 that we emit (about 55%), the rest builds up in the atmosphere (Ballantyne et al., 2012). As a result, GHG levels in the atmosphere today are higher than they have been for at least the last 800,000 years (NRC, 2010), and are rising at a rate ten times faster than the last deglaciation (Shakun et al., 2012).

It has been known since 1861 that these GHGs trap heat from the sun (Tyndall, 1861). We are now certain that over the last century or so we have changed the global climate by emitting GHGs (IPCC, 2007).

We have already seen some of the effects. It is almost certain that neither the drought in Texas (2011) nor the heatwave in Russia (2010) would have happened without recent changes in global climate (Rupp et al., 2012; Hansen et al., 2011).

Climate today

Looking at our climate situation today reveals some interesting – and troubling – changes to current local and global climatic conditions.

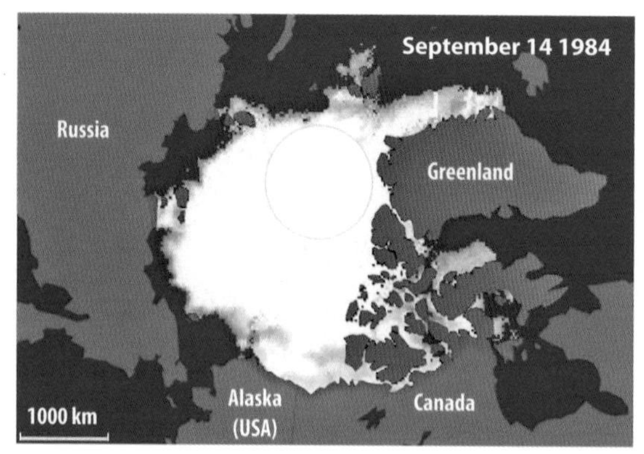

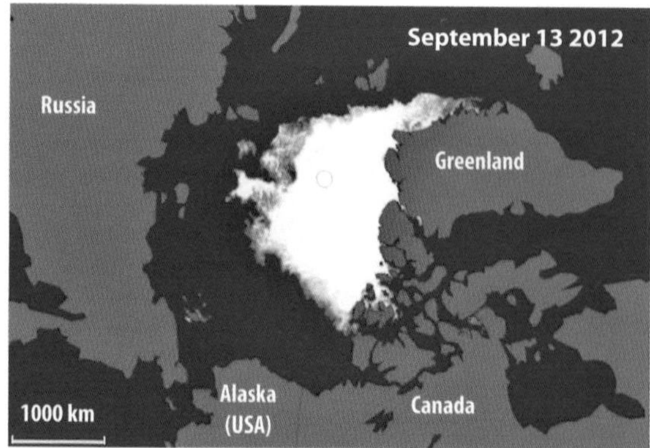

Figure 2.1: The difference in Arctic sea ice coverage during the 'summer minimum' between 1984 and 2012 (biggest ever recorded melt to date). Based on satellite data; adapted from NASA (2012).

New norms

Global average temperature has increased by about 0.8°C since pre-industrial times (Hansen, 2010). Each of the last three decades has been warmer than the previous one and warmer than any other on record since 1850. All ten of the hottest years on record have occurred since 1998 (Met Office, 2011b). The seasons are changing – spring is coming earlier and autumn is appearing later (Parmesan and Yohe, 2002; Richardson et al., 2013).

Oceans have been warming and becoming more acidic (as they absorb some of the CO_2 from the atmosphere). Sea levels are currently rising at about 3cm per decade, largely due to the fact that as water warms, its volume increases (Church, 2011).

The Arctic is warming twice as fast as the rest of the globe (Lemos and Clausen, 2009). As a result, Arctic sea ice is melting more and more every summer (see figure 2.1). The six biggest sea ice melts occurred in the last six years (2007-12 inclusive) (NSIDC, 2012) (see figure 2.2). Parts of both Greenland and Antarctica are losing ice, though less than in the Arctic and more slowly (World Bank, 2012). These changes are new and faster than climate models have predicted (Allison et al., 2009).

New extremes

Heatwaves have been getting hotter and have occurred more often. Local temperatures during the heatwaves in Europe (2003) and in Russia (2010) were much higher than 'extremes' for these places over the last 510 years (Shearer and Rood, 2011) – see figure 2.3. Hot days and nights have become more frequent (World Bank, 2012; IPCC, 2012).

The water cycle and weather systems are also changing. Warmer conditions mean more water evaporates and is held in the atmosphere (Coumou and Rahmstorf, 2012). Water in the atmosphere is the fuel of weather systems and so more water in the atmosphere can make these systems more intense (Meehl et al., 2007). For instance, there have been longer, more intense droughts in some places (Dai, 2012; IPCC, 2012), and more intense downpours of rain in others (McMullen, 2009; IPCC, 2012).

Climate tomorrow

The global community is committed to keeping global warming below 2°C to prevent dangerous climate change, with many countries pledging to cut emissions. Even a warming of 2°C would mean severe changes to the world in which we live. Many small island nations are calling for a limit of 1.5°C to be supported (World Bank, 2012), and evidence now suggests that 2°C is actually likely to be the threshold between "dangerous and 'extremely dangerous' climate change" (Anderson and Bows, 2010).

But annual GHG emissions have continued to increase – about 3.1% every year since 2000 (Peters et al., 2013). Present emissions trends (even with current pledges to cut emissions) put the world on a

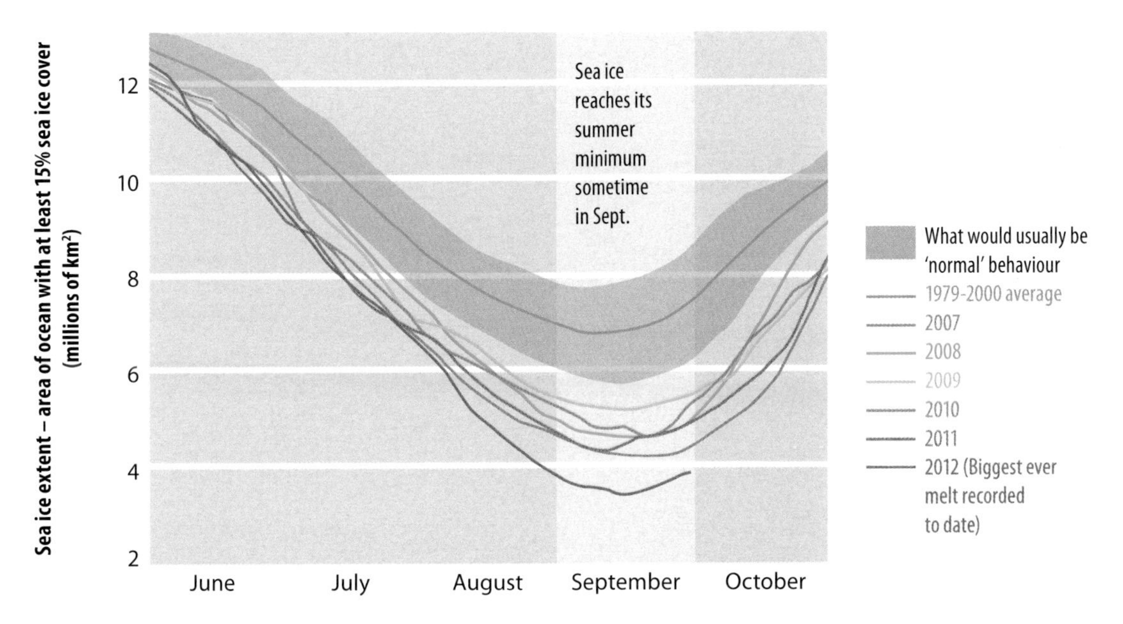

Figure 2.2: Arctic sea ice melt. The six years with the 'biggest melts' are shown relative to the average over 1979-2000 – what would usually be classified as 'normal' behaviour. Source: National Snow and Ice Data Centre (NSIDC); adapted from World Bank (2012).

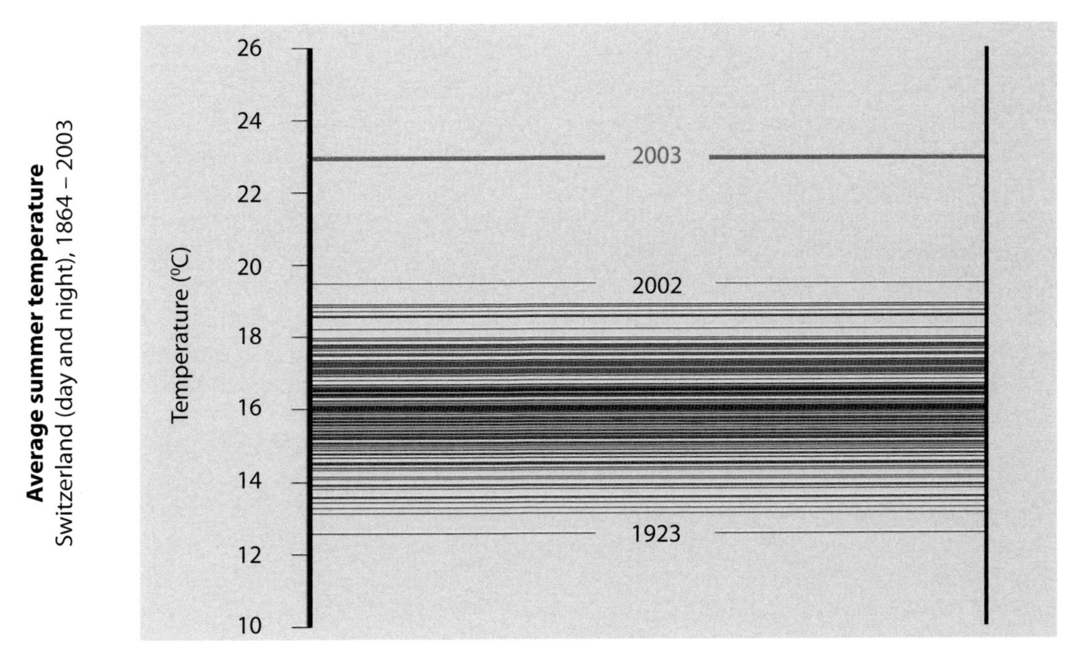

Figure 2.3: The average summer temperature during the European heatwave in 2003 relative to other years (where every line represents the average summer temperature in one year) shows how much higher it was than normal. Adapted from Schär et al. (2004).

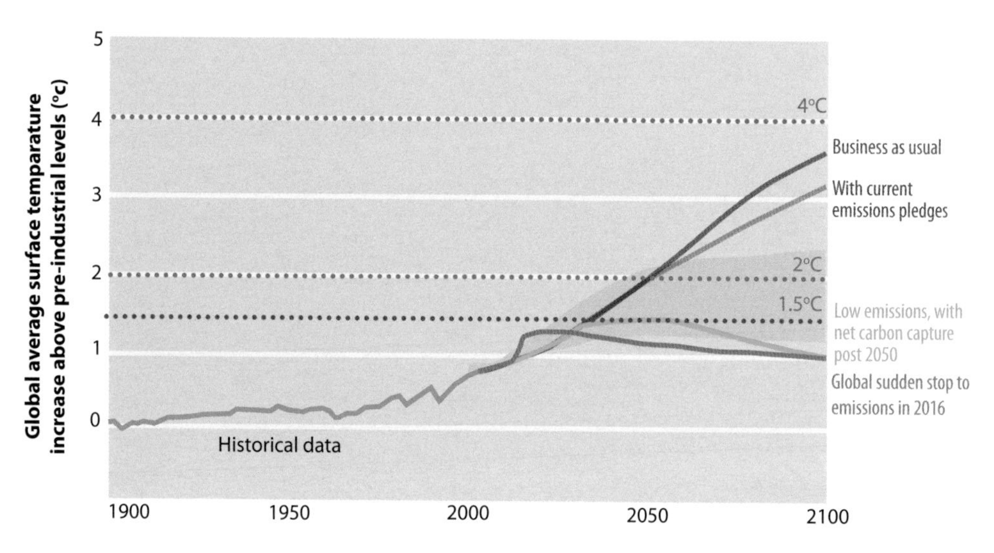

Figure 2.4: Temperature changes expected under different emissions scenarios. Adapted from World Bank (2012).

course towards an almost certain temperature rise of over 3°C by the end of the century (see figure 2.4). There is a 1-in-5 chance that continuing as we are would result in a world 4°C hotter by 2100, with even higher temperatures of over 6°C likely beyond then (World Bank, 2012).

According to the president of the World Bank (ibid.):

"The 4°C scenarios are devastating: the inundation of coastal cities; increasing risks for food production potentially leading to higher malnutrition rates; many dry regions becoming dryer, wet regions wetter; unprecedented heat waves in many regions, especially in the tropics; substantially exacerbated water scarcity in many regions; increased frequency of high-intensity tropical cyclones; and irreversible loss of biodiversity, including coral reef systems."

It is unlikely we will be able to adapt to such a world:

"There is a widespread view that a 4°C future is incompatible with an organised global community, is likely to be beyond 'adaptation', is devastating to the majority of eco-systems and has a high probability of not being stable."

Kevin Anderson, former Director of the Tyndall Centre, UK (Anderson, 2012).

A sliding scale

Many impacts work on a sliding scale – as temperatures increase, the 'norms' and 'extremes' change, and the effects become worse. A 4°C warmer world would make it possible for oceans to acidify to the point of dissolving coral reefs (World Bank, 2012), and for sea levels to rise and flood over 150 million people each year (Met Office, 2011a). Very hot days (5-10°C hotter than the current hottest days) would be much more frequent, and droughts, floods and hurricanes would likely be much more commonplace. All this would have massive impacts on the basic necessities of food, clean water, health and shelter for many across the globe (ibid.). As temperatures increase, the severity of these impacts increases.

A bumpy ride?

Perhaps even more concerning is the possibility that long-term and cascading changes would occur, making climate change much worse, much faster:

- Melting permafrost as a result of warming would mean huge releases of methane (CH_4), a powerful greenhouse gas that would contribute to warming even further (Schuur et al., 2008).
- Acidification of the oceans and the death of parts or all of the Amazon rainforest because of warming would change these systems from those that capture CO_2 to those that emit CO_2, increasing the levels of GHGs in the atmosphere (die-back in the Amazon due to localised droughts in 2005 and 2010 – both 'one-in-a-hundred-year events' – released more CO_2 than

the whole Amazon usually captures in a year (Lewis, 2011)).

- The melting of ice sheets in Antarctica and Greenland would mean over 10 metres of sea level rise, with New York, London and Taiwan under water (World Bank, 2012; McCandless, 2010).

It is not *certain* that these events will occur but, as GHG emissions continue and the global average temperature rises, the risk that they will occur increases.

In the world we are currently on course for, areas of the globe will almost certainly be completely uninhabitable, with huge ramifications on a global scale. There will be devastating worldwide impacts on the natural systems which support **all** of us. These changes would not be short-term, and would likely commit us to a worsening situation over the coming centuries.

Though there are many complex factors involved, we know that the major driver of these changes is our GHG emissions. This means that global reduction, and eventual elimination, of GHG emitting activities is necessary to change our course.

2.1.3 Planetary boundaries

Reducing GHG emissions is one of a number of major environmental requirements for global sustainability. It does not make sense to 'solve' climate change at the expense of other equally important problems. An interdisciplinary group of researchers asked:

"What are the non-negotiable planetary preconditions that humanity needs to respect in order to avoid the risk of deleterious or even catastrophic environmental change at continental to global scales?"
(Rockström et al., 2009).

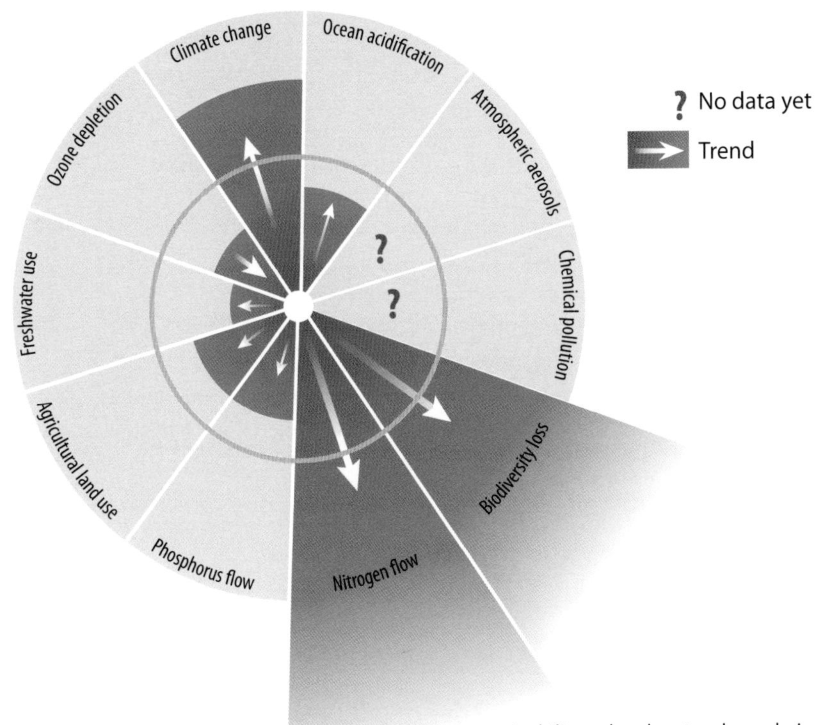

Figure 2.5: Illustration of the environmental requirements for global sustainability – the planetary boundaries. The grey circle represents the 'safe operating space'. Shaded areas represent the current state and trends of the environmental problems relative to their 'safe' boundaries. Adapted from Rockström (2010).

Environmental problem	Why it's important	Causes	Current situation and trends
Climate change	See *2.1.2 Climate change*.	Continual GHG emissions build up in the atmosphere.	The 'safe' boundary is already exceeded and the problem is escalating.
Ocean acidification	We rely on the oceans for food; they regulate the climate by absorbing CO_2 and play an important part in regulating weather systems.	Oceans take in some of the excess CO_2 in the atmosphere that has built up through continuous GHG emissions.	Still within the boundary, but the situation is deteriorating.
Ozone depletion	The ozone protects us from harmful radiation from the sun.	Chemicals such as CFCs emitted into the atmosphere.	Within the boundary and improving.
Nitrogen flow	Plants need nitrogen to grow, but high levels in freshwater leads to pollution and kills aquatic life.	Largely due to overuse of fertilisers for agriculture.	The 'safe' boundary is already greatly exceeded and the situation is deteriorating.
Phosphorus flow	Plants need phosphorus to grow, but high levels in freshwater leads to pollution and kills aquatic life.	Largely due to overuse of fertilisers for agriculture.	Still within the boundary, but deteriorating.
Freshwater use	We require freshwater to live – for drinking and watering the crops that feed us.	Multiple causes including overuse of groundwater stores and large demand for water in dry areas.	Still within the 'safe' boundary globally, but deteriorating. In many places the 'safe' boundary at the local level is exceeded.
Agricultural land use	Land not used for agriculture supports biodiversity and acts as a climate regulator, absorbing CO_2.	Cutting down forests and converting land to agricultural use (cropland and intensively grazed grassland).	Still within the boundary, but deteriorating.
Biodiversity loss	High biodiversity means a far more resilient, adaptive ecosystem – often the ecosystems on which we rely.	Overfishing in oceans and invasive species on land and in oceans – alongside other problems listed here – contribute to biodiversity loss.	The 'safe' boundary is already greatly exceeded and the situation is still deteriorating rapidly.
Chemical pollution	Pollution has a detrimental effect on ecosystems and can impact us by working its way up the food chain, leading to health risks.	Multiple causes including agricultural run-off, industrial spillages and lack of stringent controls on chemical use.	Not yet measured.
Atmospheric aerosols	These change weather systems, particularly rainfall and monsoon patterns, leading to floods and droughts.	Multiple causes.	Not yet measured.

Table 2.1: Each environmental problem with details of the current situation, trends and the main causes of each problem, listed (Rockström et al., 2009; Raworth, 2012).

They identified several key environmental problems and suggested safety limits for each. They called these limits 'planetary boundaries' and proposed that together these boundaries defined 'the safe operating space for humanity' – preconditions for global sustainable development (or human existence). Though there are many ways of assessing the relative importance of environmental (and other) problems, the critical point here is that identifying multiple issues helps to avoid selecting solutions that solve one problem at the expense of others. Instead, solutions that help address multiple issues are promoted.

Table 2.1 and figure 2.5 describe the problems in more detail.

Of these boundaries three are already exceeded, four are deteriorating, two have not yet been quantified and only one – ozone depletion – is improving (Rockström et al., 2009). In general, our actions are contributing to the continual worsening of a broad range of dangerous environmental problems, not just climate change (Rockström and Klum, 2012).

2.1.4 Future generations

Most people feel that it is right to take the interests of future generations into account when discussing such large-scale environmental problems, yet this is difficult to implement – our descendants are not actually here to argue for their rights.

A useful frame was set by the Brundtland Report of 1987, which proposed we 'provide for our own needs without compromising the needs of future generations'. It put on the agenda the key idea of a balance of interests between present and future generations. The idea was enshrined in the United Nations Framework Convention on Climate Change (UNFCCC) in 1992, setting out the case for global action on climate change.

Yet we remain for the most part 'future blind'. Traditional economics supports this idea by assuming the ability of future generations to solve environmental problems is best served simply by maximising economic growth in the present, and handing on the material benefits. The argument is that by generating wealth now, it will be cheaper to act in the future given that we hand future generations 'better tools' for dealing with problems. Future costs are progressively 'discounted' at a rate of about 5% a year, so are assumed to diminish to almost nothing (Beckerman, 1995; Nordhaus, 2007).

Other economists, however, have urged that the interests of future generations should be treated as having the same value as our own, prompting a much more precautionary approach. For example, recent work by Stern (2009) estimates that an investment of 2% of UK gross domestic product (GDP) now could be sufficient to prevent future costs in the region of 20% of GDP. The argument is that given less optimistic (and arguably more realistic) assumptions about what happens in the future, it is better to act earlier rather than later.

Our responsibility to future generations must also extend beyond economics: **it is unethical to treat fundamental needs in the future as equivalent to our lifestyle preferences today**. The evidence for high risks of extremely grave outcomes cannot be ignored. If global mitigation is unsuccessful, the worst-case outcomes include widespread state collapse, breakdown of the international order, hundreds of millions of environmental refugees, climate wars, failure of the ecosystems on which we depend and mass extinction (World Bank, 2012; Barnett and Adger, 2007; Sachs, 2007). It would be absurd to treat these outcomes simply in terms of economic costs that, however high, would constitute "a serious underestimate of infinity" (Toman, 1998).

Making relatively small – though not insubstantial – changes now will help protect future generations from a situation that, at best, would mean making much larger changes and, at worst, would simply be unliveable (see figure 2.6 below). By making falsely optimistic assumptions about the future now – economically, and in terms of the risks of climate change – we are not looking out for future generations.

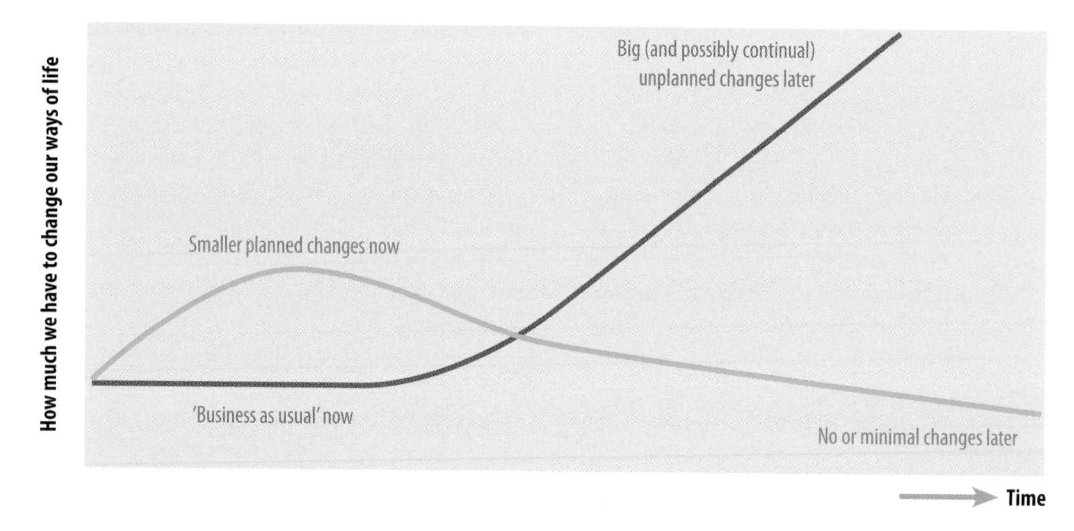

Figure 2.6: Illustration showing that choosing relatively small and planned changes now can avoid potentially much larger and unplanned changes to our ways of life later.

2.2 The situation in the long-industrialised West

© Joanna Wright

Though we have benefited hugely from industrialisation, the Western world has also created many problems for others as well as for ourselves in the process. In many cases, we have externalised the effects of our actions both economically and physically – by not counting the broader environmental impact of the actions we take, and by literally 'offloading' detrimental impacts elsewhere. In other cases, we have simply been looking in the wrong direction – favouring and fostering economic development over our happiness and wellbeing, for example.

As part of a global system, we rely on stable supplies of energy, goods and services from around the world to satisfy our 'needs' – meaning any global problem is also a local one. We fail to recognise that continual growth is not possible in a finite world, and are beginning to see the effects on our local energy supply, economy and our happiness as individuals and as societies.

2.2.1 Energy supplies

Climate change and environmental degradation are not the only drivers for a transition away from fossil fuels. Our fossil fuel based economies are being halted by the immovable facts of geology. For the first time in our history, and just as demand is exploding across the globe, humanity is close to no longer being able to increase annual energy production using fossil fuels. Despite the accelerating energy demand, global rates of 'conventional' oil and gas production are heading towards an inevitable plateau beyond which they must go into decline, with the remaining fossil resources being dirtier, harder and considerably more expensive to extract.

Consumption of oil has risen to nearly 33 billion barrels a year (some 90 million barrels per day) and the price has increased tenfold over the last century. This is mainly because sources of cheap, 'easy' oil are dwindling rapidly (Johnson et al., 2012). In the 1930s, burning oil produced about 100 times the energy used to extract it. But, as oil has become harder to get at, the amount of energy used to extract it has increased. By the 1970s, burning oil produced only 30 times the energy needed to extract it. Today, most new oil discoveries produce only ten times the energy we use to get it out of the ground (Morgan, 2013).

The peaking of global oil and gas supplies offers one clear reason to move beyond fossil fuel dependency: not because the supply will run out in the near future, but because the escalating prices will cause increasing turmoil in the economies (and societies) that still depend heavily on them.

Fracking: an answer for the UK?

Here in the UK, hydraulic fracturing (or 'fracking') is proving highly controversial. It involves inserting a mix of chemicals under high pressure into an area underground to release gas trapped in shale rock. It is unlikely to offer a lasting solution for our energy needs. Estimated yields from UK hydraulic fracturing fields are 150 billion cubic metres, equivalent to 1,470 TWh per year, or around a single year's primary energy production for the UK (Richards, 2012). Conventional gas fields decline relatively slowly whereas shale gas declines very rapidly, as pressure within the earth closes up the fissures being exploited. There is also concern over earthquakes, pollution of water supplies and the effects on wildlife.

The UK is now at a critical crossroads, as a significant amount of our current generation capacity is due for retirement within the next ten years. Strategic thinking is vital now to avoid panicked choices that will lock the UK into a problematic energy path for the future. Any investment in new generation plant infrastructure must take full account of the longevity of the fuel supply, the cost of extracting fuel and producing energy, as well as the potential fuel price rises that may occur during its design life.

In 2005, the UK became – once again – a net energy importer (DECC, 2009). Whilst increasing fossil fuel imports can substitute for falling domestic production in the immediate term, this is not a secure long-term solution due to global geological constraints. As supplies struggle to keep up with demand now, global oil and gas prices look set to rise, affecting our security of supply and damaging the UK's economy while potentially contributing to fuel poverty.

2.2.2 The economic crisis

Communities across the globe have been struggling to adjust to a new era of profound and abrupt change. The spiralling energy prices and financial crash of 2007 not only revealed an enormous burden of hidden debt, but also led to the largest and deepest period of economic turmoil in generations. In 2013, the effects of this collapse continue to roll on – from Spain to Cyprus there is a fresh crisis on an almost monthly basis. Governments and communities now face the twin questions of what must be done to prevent this happening again, and how. These challenges are set against increasing energy shortages and price volatility. Are governments and communities able to guide their economies through a process of debt reduction and economic regeneration?

Origins

Back in 1964, bank managers were renowned for being prudent – UK household debt was running at around 14% of GDP. However, following the deregulation of the 1980s and 1990s, it increased to 80% (Elliot and Atkinson, 2012). Cheap, deregulated finance not only enabled us UK consumers and producers to live beyond our personal finances, but also beyond our fair share of global resources and the means of the environment, all to provide for our needs and deal with our wastes.

Market rules were set well before we were concerned about either climate change or oil depletion. Consequently, they are carbon-blind. The influential 18th century economist Adam Smith and his 'invisible hand of market mechanisms' still fumbles blindly, guiding choices that might stack up economically but do not reflect the true cost of the damage to natural ecosystems – a cost that neither the producer nor consumer pays.

Yet even with a significant part of the *real* costs not included, the UK's economic plan has still not really delivered. As industry in the UK scaled down, debt-driven consumer spending increasingly became the engine powering the UK economy. Initially, this appeared to be working – with a vibrant high street economy, Britain was once again a nation of shopkeepers, albeit mostly large corporate chain stores. But following the collapse of cheap credit, the economic approach was brought to a juddering halt. To make matters worse, growth in lower-cost online sales direct from overseas manufacturers have caused the high street chain store economy to falter.

Genuine recovery will require a new plan for going forward. We are beginning to recognise that adopting new attitudes and approaches to energy and environment must form a fundamental part of this shift – we can no longer expect continual growth on a planet with finite resources and environmental limits.

2.2.3 Wellbeing

Growth in fossil fuelled consumer culture isn't just wrecking the wellbeing of the planet – the tendency to base our identities on money, possessions or appearance is also seriously affecting our own health and happiness.

The practice of acquiring material possessions in excess of needs as a way of displaying status is as old as civilisation itself. However, the rise of abundant cheap fossil fuels has provided the means for this conspicuous consumption to be globally flaunted – a situation unique in human history.

Driven by powerful advertising and easy credit, we seek ever-higher levels of material consumption in the belief that this will lead to increased respect from our peers and a better, happier life. We are acquiring more than any human society has ever acquired before; shouldn't we be happier than ever before?

Clearly, below a crucial threshold we will be unhappy – when we don't have enough to eat or

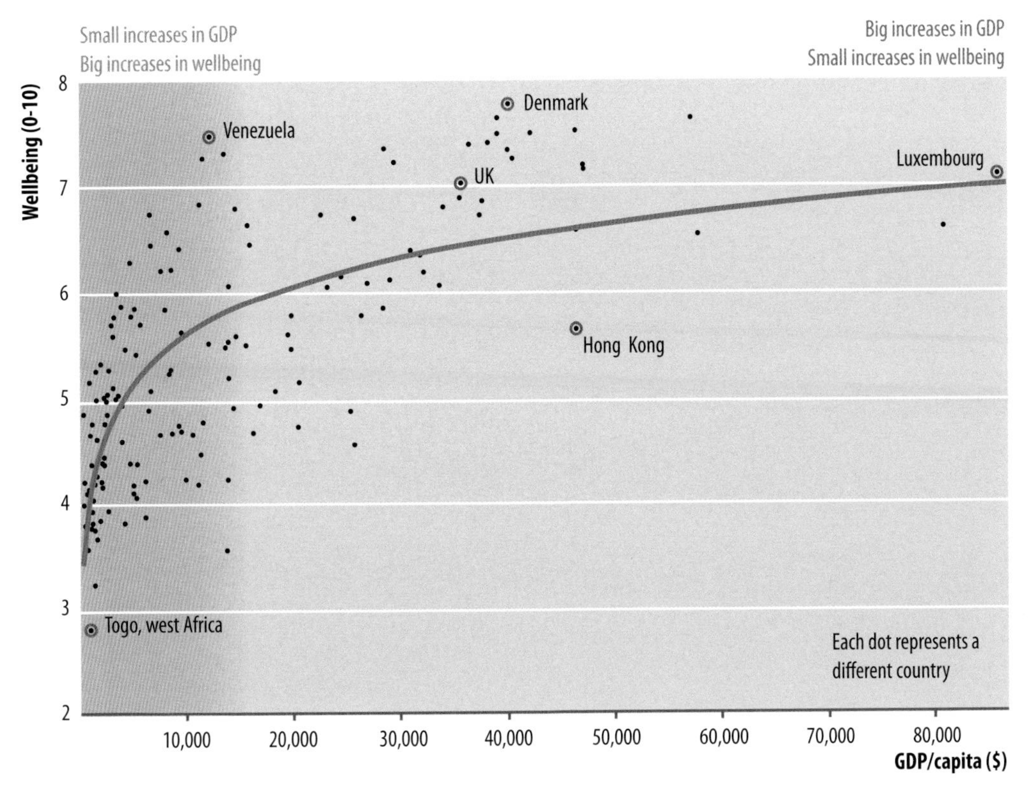

Figure 2.7: Wellbeing (as rated by individuals in a survey on life satisfaction) versus gross domestic product (GDP) per capita – consuming more doesn't necessarily lead to greater wellbeing past a certain point.
Based on data from Abdallah et al. (2012).

when we can't keep our children or ourselves warm, sheltered and clothed. But a growing body of research reveals that even far above this basic level, using or having extra energy or materials is not necessarily bringing higher levels of happiness or wellbeing (see figure 2.7). Significantly, only around 10% of the variation in subjective happiness observed in Western populations is attributable to differences in material circumstances such as energy use, income and possessions (Lyubormirsky et al., 2005).

People tend to adapt relatively quickly to increases in material consumption, soon returning to their prior levels of happiness (Abdallah et al., 2006 and 2009; Thompson et al., 2007). Even more surprisingly, the richer a nation gets (once it moves beyond 'enough'), the more unhappy and unhealthy its people can become – though some of this is due to the inequality in these situations, rather than absolute wealth.

Inequality contributes to a large number of social problems that influence the wellbeing of those at the 'top of the pile' as well as at the 'bottom' – poor health, higher levels of violence and drug abuse, and lack of trust amongst others (Wilkinson and Pickett, 2009). Inequality is on the rise in the UK. By 2007-8, the UK had reached the highest level of income inequality since shortly after the Second World War (Hills et al., 2010) – the total household wealth of the richest 10% is over 100 times that of the poorest 10% (ibid.). The average chief executive officer (CEO) pay was 47 times that of the average worker in 1998; by 2008 it was 128 times greater (Peston, 2009).

Today we are using far more energy than we actually need, while obesity and isolation, personal debt and physical inactivity are reaching record levels.

2.3 What does this mean for the UK?

Obviously, no single nation on its own can solve the climate problem. It has to be a collective effort. The United Nations Framework Convention on Climate Change (UNFCCC) was created in 1992 to address this problem, and it committed signatories to take steps to avoid 'dangerous climate change' (agreed as a global temperature rise of over 2°C relative to pre-industrial levels). Like the subsequent Kyoto Protocol (UNFCCC, 1998), it requires each nation to make an appropriate contribution.

But what is the UK's appropriate contribution? By the standards of international climate diplomacy, the UK has been something of a leader. It has set a number of binding emissions targets relative to 1990 when GHG emissions were 778 million tonnes of carbon dioxide equivalent ($MtCO_2e$). The targets are:

- The 2012 Kyoto Protocol target of 12.5% reduction to 681 $MtCO_2e$ (UNFCCC, 1998).
- An 'interim target' of 34% reduction to 513 $MtCO_2e$ (DECC, 2011).
- A long-term 2050 target of 80% reduction to 157 $MtCO_2e$ set out in the UK Climate Change Act 2008 (HM Government, 2008).

These substantial reduction targets are backed by UK law. The Kyoto Protocol target was already achieved by 2000, and the general trend has been steadily downwards, broadly in line with the long-term target (see figure 2.8).

But is this consistent with global requirements? Science has moved on since 1992, and it has become clear that it is not the end point (or target) of emissions reduction that constrains global temperatures, but *the total quantity of emissions* along the way (Messner et al., 2010). Which begs the question, do current targets keep us within this new constraint?

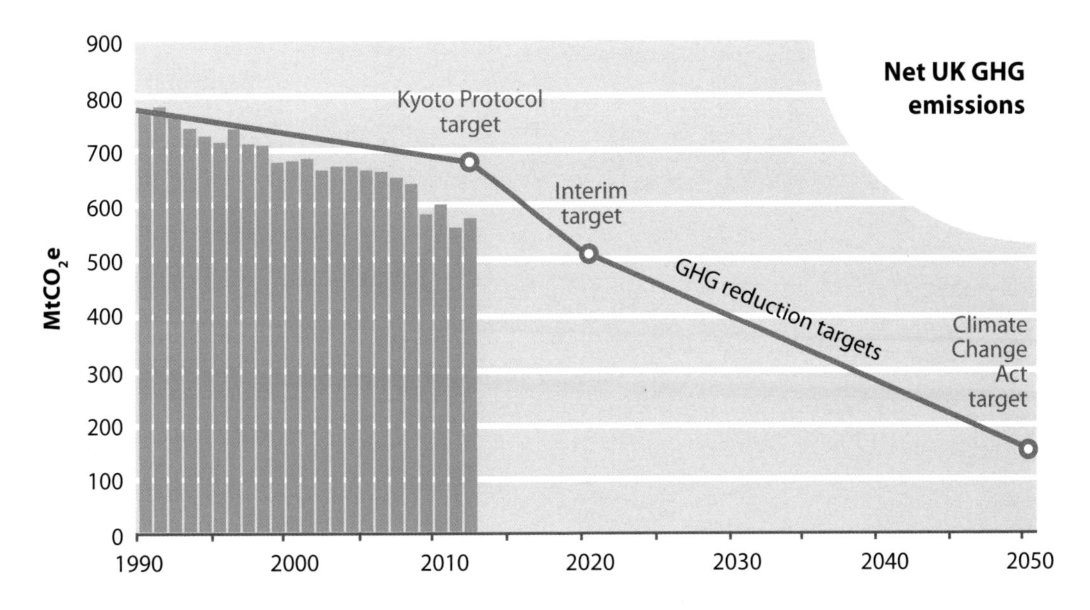

Figure 2.8: Annual production emissions of the UK ($MtCO_2e$) from 1990 onwards (not including international aviation and shipping), showing progress alongside our internationally agreed emissions reductions targets. Adapted from Beales (2013).

Furthermore, as a long industrialised nation we have contributed significantly to global emissions over the last 150 years or so, enabling progress and development and getting us to what we are today – a wealthy Western nation. Do current targets show that we are approaching the challenge of creating a sustainable future both fairly and equitably?

2.3.1 Our carbon budget

There is widespread acceptance that collective global policy should not permit an average temperature rise greater than 2°C (Messner et al., 2010). To achieve this, a cumulative 'carbon budget' for the world can be set – defining how much GHGs can be emitted in total. The world (mostly the Western world) has, of course, already 'spent' a large proportion of this budget. A global cumulative carbon budget measures 'what is left' at a particular date, and decreases every year we keep emitting GHGs.

There is, however, much uncertainty about what size a global carbon budget should be if it is to give us a good chance of avoiding a 2°C global average temperature rise.

Indeed, what constitutes a 'good chance' is difficult to define.

One study (Meinshausen et al., 2009) calculates global cumulative GHG budgets between 2000 and 2050. Global GHG emissions between 2000 and 2009 alone were 400 gigatonnes (Gt) CO_2e (FoE, 2010), meaning we have already 'spent' a large proportion of what is available to us until 2050. This means that between 2010 and 2050 a global carbon budget of:

- 950 $GtCO_2e$ would give us an 80% chance of avoiding a 2°C global temperature rise.
- 1,100 $GtCO_2e$ would give us a 75% chance.
- 1,280 $GtCO_2e$ would give us a 67% chance.
- 1,600 $GtCO_2e$ would give us only a 50-50 chance.

A defined global budget can then be 'shared out' between nations according to their population, meaning larger nations have a larger budget. A globally equitable per capita budget of this kind is the most likely basis for the necessary post-Kyoto Protocol treaty required for successful global decarbonisation (Messner et al., 2010).

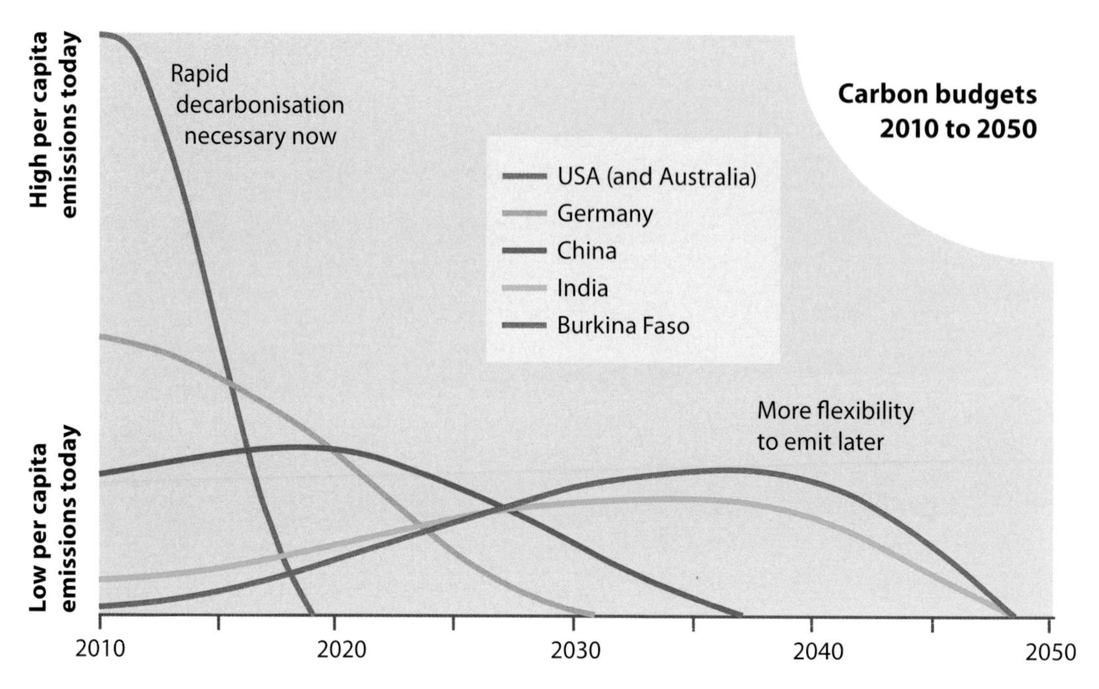

Figure 2.9: Examples of the difference that carbon budgets make to the decarbonisation trajectories of countries that currently have very high emissions and those which have low emissions. Adapted from Schellnhuber, 2009.

An important consequence of equal per capita budget allocation is that countries with high per capita emissions must reduce very quickly to stay within their budget, while those with low per capita emissions have greater flexibility, and are free even to increase their emissions if they consider it necessary. This is illustrated in figure 2.9.

Are we keeping within our budget?

Assuming an average global population of roughly 8 billion, and an average UK population of 70 million between now and 2050, the UK's share of the global budget between 2010 and 2050 would be about:

- 8,400 $MtCO_2e$ (80% chance of avoiding a 2°C global average temperature rise).
- 9,600 $MtCO_2e$ (75% chance).
- 11,200 $MtCO_2e$ (67% chance).
- 14,000 $MtCO_2e$ (50% chance).

This covers all that we could 'spend' (or emit) between 2010 and 2050.

As the UK government has already published a series of legally binding carbon budgets up to 2028, and further emissions reductions to 2050, we can calculate roughly how much carbon we will 'spend' if we meet all our targets.

Using data for UK GHG emissions from 2000-12 (DECC, 2013b; and DECC, 2013a) and a projection of GHG emissions in line with current policy targets, we find that the UK will emit about 15,800 $MtCO_2e$ (16,000 $MtCO_2e$ including emissions from international aviation and shipping – currently not counted under the Kyoto Protocol) by 2050 – well over the amount for even a 50% chance of avoiding the 2°C limit.

Such a budget would not be acceptable in international negotiations, especially in view of the fact that most of the present atmospheric GHGs were generated by wealthy countries like the UK during their development process. In some sense, such countries have already exhausted their 'moral budget' – having emitted far more than their 'fair share' over the years since the industrial revolution – and should perhaps shoulder this 'historical responsibility'.

From this perspective even 8,400 $MtCO_2e$ might be considered generous (Wei et al., 2012). For more discussion on what effect taking responsibility for our historical emissions has on the UK's 'fair share' of a global carbon budget, see *3.8.1 ZCB and the UK's carbon budget*.

2.3.2 The physics-politics gap

Physical problems have physical solutions and no amount of talking will make them go away. This is not to say that talking is not important; it is essential. But it is best to get the physics right first.

Virtually everybody agrees that rapid decarbonisation is the cornerstone of any solution to climate change, and we have adequate ways of measuring how much decarbonisation is required, plus how fast it is required.

However, if we analyse these physical requirements and work out a physically credible plan based on our scientific knowledge of the situation, we find it does not fit comfortably into the frame of normal politics and economics. On the other hand, if we work out a plan that does fit the politics, we find it does not meet the physical requirements. In fact, a huge gulf between what is physically demanded by science and what is seen as politically possible is revealed. This is reflected in the difference between our projected emissions 'spend' above (15,800 $MtCO_2e$), and the UK's portion of the global carbon budget in line with a good (80%) chance of avoiding a global temperature rise of 2°C (8,400 $MtCO_2e$). That's a difference of 7,400 $MtCO_2e$.

We can call this the 'physics-politics gap', as illustrated in figure 2.10.

Most current efforts attempt to build bridges from the now, working forwards within current political, economic and social boundaries to try and meet the challenge of rapid decarbonisation. There are plenty of 'half bridges' built on foundations in the politically realistic perspective, none of which quite reach where we need to go from the physically realistic perspective.

Another approach is to instead ask, 'what is the

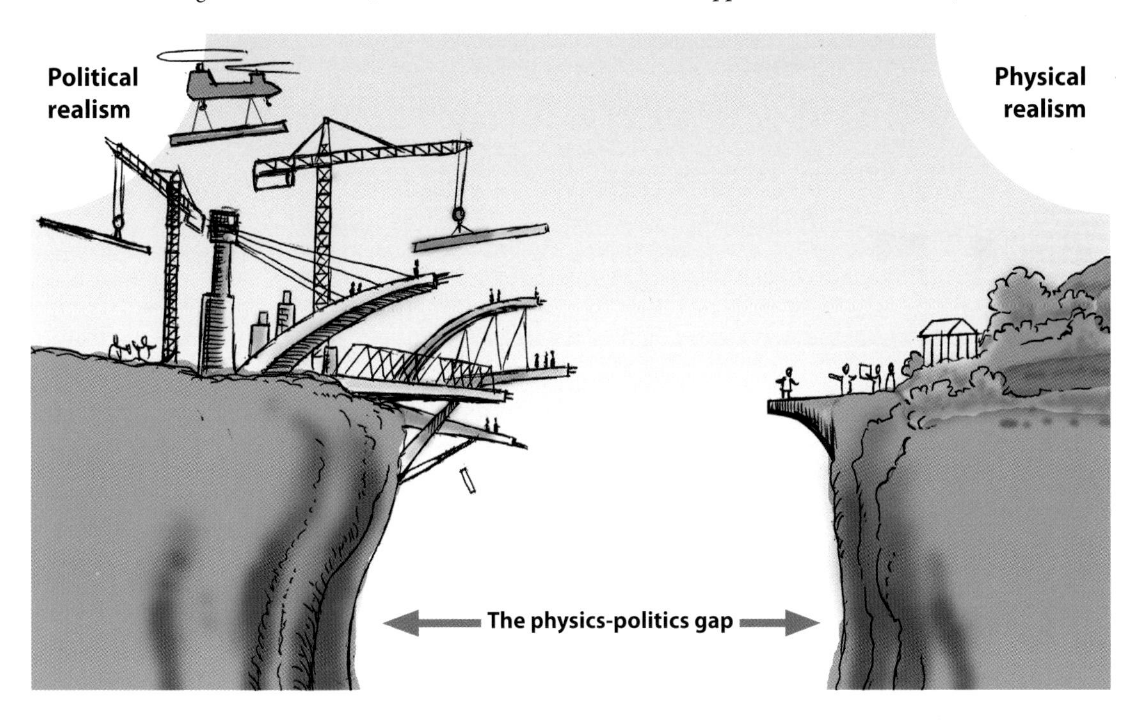

Figure 2.10: An illustration of the physics-politics gap and efforts to try to bridge it from the politically realistic side. A physically realistic perspective sits on the other side of the gap, denoting where we need to be to meet the physical requirements of the problem.

end point?' A physically realistic perspective based on this line of question shows us where we need to get to in order to successfully meet the challenge of climate change. We can explore the possibilities for physically realistic worlds and consider what needs to change (from lifestyles, to infrastructure, to politics and economics) for us to get there, plus how fast we need to change, and the alternative routes that we can take.

Once we have worked out where we need to get to, we can work backwards to find out how we get there. Zero Carbon Britain focuses on the questions involved in this process and sets out such a physically realistic scenario – laying foundations on the 'right' side of the physics-politics gap.

Chapter 3

Our scenario: Rethinking the Future

3.1 About our scenario

The great thing about the future is that anything can happen. Many radical changes throughout history have shown us that we are bold, innovative, creative and often surprising individuals and societies.

Trying to figure out 'where to go from here' can be restrictive, as it means operating within the systems and constraints we recognise and know. But figuring out 'where we want to end up' is both exciting and overwhelming in equal measure.

It is necessary here at the start then to reduce the very large – potentially infinite – range of future scenarios we could construct. We can do this by giving ourselves some aims to work towards and some rules by which to play. We must also state our assumptions – we're not modelling the entire world here, so we need to figure out on which basis we are setting our scene.

3.1.1 Aims

The first aim is obvious – to make our contribution to addressing climate change. To do this we must become 'net zero carbon', since any remaining emissions (no matter how small) that are not balanced with equivalent carbon capture methods will eventually add up, contributing to atmospheric greenhouse gases (GHGs). This must be sustainable in the long-term, and the transition rapid enough for us to maintain a carbon budget that gives a good chance of keeping global average temperature rise below 2°C. Our net zero carbon scenario is therefore set in 2030.

We also want to make sure that decarbonising the UK in these ways does not mean living in a cave and eating bugs off the walls. Our wellbeing – physical and mental – and that of the local environment is important, and so we also need to:

- Keep the lights on and keep everyone warm, providing enough energy to meet demand at all times.
- Make sure we all eat enough, and eat well.
- Keep a decent standard of living, with the benefits of a modern society.
- Support biodiversity – making space for the natural world we rely on.
- Look at how to help adapt to a changing climate – building resilience into our systems to be able to respond to the foreseen and unforeseen effects of climate change.
- Weigh up the costs and benefits (not just monetarily) of our options.

Although living in the UK will be different in our scenario, we create a scenario that represents a positive future – one that inspires change.

3.1.2 Rules

Rules are born out of the values we hold as individuals and societies. They guide us when making decisions and make it easier for us to check that what we are doing is fair, and that we are meeting our aims.

We have made the following list of rules to guide us in creating our scenario.

When counting GHG emissions we:

- Must include all the different GHGs as recorded by the United Nations Framework Convention on Climate Change (UNFCCC): carbon dioxide (CO_2), methane (CH_4), nitrous oxide (N_2O) and others. Since all of these GHG emissions contribute to climate change, we have to reduce them all. To make this easier, we measure them all in 'carbon-dioxide-equivalent' (CO_2e) – the equivalent impact in terms of CO_2 of each gas over the standard 100-year timeframe. For example, methane is 21 times more powerful as a GHG than CO_2, so 1 tonne (t) of methane equals 21 tCO_2e (ONS, 2012). For the CO_2e of other GHGs, see End notes.
- Count carbon on a 'production' basis. This means we take into account all the GHGs emitted within the borders of the UK. We also include those from our share of international aviation and shipping (not currently included in UNFCCC totals). Counting carbon from a 'consumption' basis is discussed in *3.10.3 Carbon omissions.*
- Start with the UK GHG emissions in 2010 (DECC, 2013). These emissions (UNFCCC and international aviation and shipping) come to a net total of about 628 $MtCO_2e$. We then calculate the additional impact of aviation, as GHGs emitted higher in the atmosphere may have a greater warming effect (Lee, 2010). This brings the net effect of the UK's actions in 2010 to about 648 $MtCO_2e$.

What do we mean when we talk about 'emissions' and 'zero carbon'?

In this report, we talk about both carbon emissions and carbon capture. These two things are usually combined in UK GHG emissions accounts.

In 2010, the UK actually *emitted* 652.1 $MtCO_2e$ including international aviation and shipping. But in the same year natural systems in the UK *captured* 23.8 $MtCO_2e$ of carbon, balancing out some of our emissions. These two figures combine to give the total **net emissions** of 628.3 $MtCO_2e$ (652.1 $MtCO_2e$ *minus* 23.8 $MtCO_2e$). On top of this, we add the additional effect of aviation, getting a total of 647.5 MtCO2e – this was the UK's estimated **net effect** or **net impact** in 2010.

When we talk about **emissions** in the report – for example, 'about 82% of our emissions come from energy use' – we are referring to the first figure here: the UK emissions totalling 652.1 $MtCO_2e$ in 2010.

When we talk about **net emissions** in the scenario, we are referring to emissions minus carbon capture – net emissions in 2010 were 628.3 $MtCO_2e$.

However, when we talk about becoming *zero carbon,* we are talking about the **net effect** on the climate, including the effects of flying – the UK's net effect was equivalent to 647.5 $MtCO_2e$ in 2010.

In creating our scenario we:

- Use only technology available now and currently in use, or technologies which have been demonstrated to work. This ensures that our scenario is realistic in technological terms – we don't rely on silver bullets (promises of future developments in technology). We need to act now on climate change, and so we must present solutions that could be implemented immediately.
- Propose changes that last – there is no point looking simply at the short-term. Any solutions we propose must have the capacity to last for the rest of this century, although hardware might need replacing and maintaining over this time, of course. Some short-term measures can, however, help during the transition to a zero carbon Britain (see *3.6.3 Capturing carbon*, and *3.8.1 ZCB and the UK's carbon budget*).
- Rely on well established research wherever possible. Some areas of scientific research are not well quantified, however. Where science currently doesn't have an answer, we should be cautious, and not overextend the effect of an action.
- Supply our energy with 100% renewable technology, with no nuclear component. Even today, there is no plan for the waste from many of the UK's current nuclear power plants – it will have to be kept safe for thousands of years to come (DECC, 2011). Nuclear plants, and the hazardous waste produced, substantially increase the risk of very serious and lasting damage from natural disasters, climate-related events, or global political instabilities. Renewable energy systems do not have costly or difficult waste to manage, they don't require expensive and lengthy decommissioning processes, and they are at a much lower risk of very serious lasting damage from unpredictable future events.
- Rule out geoengineering options (see box on page 33) that are considered potentially dangerous, are only in early stages of development, or have not yet been proven to work. This leaves us with the following options:
    - Planting forests.
    - Producing biochar for soils.
    - Permanent burial of biochar or organic material ('silo storage').
    - Carbon Capture and Storage (CCS) at fossil fuel power stations or industrial plants.
    - Bio-energy Carbon Capture and Storage (BECCS) – carbon capture and storage at biomass-based power stations.
    - CO_2 air capture ('scrubbing') and storage – direct mechanical capture of CO_2 from the air.

  Public support appears highest for planting forests and producing biochar, which were perceived as more 'natural' geoengineering methods (Ipsis Mori, 2010). For these reasons, the first three options are prioritised in our scenario. Fossil fuel power coupled with CCS does not provide a solution. Not all the GHG emissions are captured from the fossil fuel plant, and, as highlighted in *2.2.1 Energy supplies,* fossil fuels reserves are becoming dirtier, harder and considerably more expensive to extract. The storage suggested for carbon captured through CCS, BECCS and CO_2 capture from the air, is usually old oil and gas fields (on land or under the sea), which must be monitored indefinitely to minimise leakage. This implies unknown costs and effective risk management long into the future, which cannot be guaranteed. Whilst abrupt leakage events might be seriously damaging to local systems (especially if the storage is underwater), diffuse leaks can be more difficult to stop and would, at least in part, reverse the mitigative effect of capturing the GHG emissions in the first place (IPCC, 2005). There are also limits to the CO_2 storage capacity of most methods, meaning that these options do not represent *alternatives* to decarbonisation, and in the long-term they would be phased out (Vaughan and Lenton, 2011).
- Do not rely on international or transitional credits. Funding the transition to zero carbon economies in less developed nations by paying so that we can emit more than our fair share of GHGs, or paying them to capture equivalent

carbon on our behalf, can be seen as a positive outcome of our inability to reduce emissions sufficiently in scale or speed. However, it is difficult to tell, without modelling the rest of the world, how many credits would be 'fair' (or indeed possible) to use, and lenient rules can lead to double counting, meaning global emissions reductions are eventually not met (UNEP, 2012). As such, we don't inherently think there is anything wrong with the purchase of international credits, if the scheme is implemented well, but choose not to rely on them in our scenario. International credits do not provide a long-term solution to GHG emissions, are not an alternative to decarbonisation and can delay the urgent need for action on climate change in long industrialised nations (ibid.).

And finally, with reference specifically to our aims, we:

- Must make sure energy supply meets energy demand, at all times. This follows on from our aim to keep the lights on and to keep people warm.
- Only rely on renewable energy sources inside the UK (including UK offshore waters). Importing energy from other countries need not be a bad idea, but it is difficult to guarantee the reliability of energy imports or to ensure that we will only take our 'fair share'. As such, we choose only to use energy we can produce at home.
- Must ensure that the food we produce feeds the UK population sufficiently and healthily. We choose not to import livestock or feed for livestock, as this has detrimental impacts elsewhere in the world (Audsley et al., 2009).
- Must not increase the area of land managed by us – we must leave wild areas and room for conservation and habitat restoration or protection. At the very least, this will mean we do not further damage local environments. Other needs of the land (aside from carbon management) must be considered – including biodiversity and human enjoyment.
- Choose solutions that help us adapt to a changing climate, where possible. Despite efforts to mitigate climate change, there are some unavoidable impacts already 'in the pipeline' (Jenkins et al., 2009). We must therefore try to make sure our scenario provides flexibility to adapt to these changes.

What is geoengineering?

The term 'geoengineering' can cover many different technologies and techniques that aim to mitigate climate change or the effects of climate change – from planting forests to capture CO_2 from the air to deploying mirrors in space to reflect the sun's rays and cool the planet. These examples describe the two main types of geoengineering – those that directly reduce levels of CO_2 in the atmosphere, and those that reduce the warming effect of increased levels of CO_2 in the atmosphere. The latter type do not address all of the impacts of climate change, however – for example, ocean acidification would continue even if temperatures were prevented from further rising (Vaughan and Lenton, 2011). A report by the Royal Society, *Geoengineering the climate* (2009), assessed many geoengineering options on their effectiveness, their 'timeliness' (how close to being technically viable and how quick to work), their potential cost and their safety. Geoengineering options vary hugely in all these areas, and also have significant governance and policy implications.

Oh, and despite the project title, we don't just model Britain; we really mean the whole of the UK. Most data are provided for the UK rather than Britain. Climate change policy must be supported by central government, and our legally binding international targets on GHG emissions are for the UK, so it makes more sense to include us all.

3.1.3 Assumptions

About the UK in our scenario:

- The population of the UK increases as per projections. This means that in 2030 there are about 70.6 million people (ONS, 2011). Our scenario must cater for this population – from energy demand through to food provision.
- Official projections assume a decrease in household occupancy in the future (Communities and Local Government, 2010) but this is uncertain enough that we assume the average household stays roughly the same as it is now, about 2.36 people per household (ONS, 2012). Since larger households use energy more efficiently (Utley and Shorrock, 2008), we should aim to actually *increase* the number of occupants in a household over the long-term (though not indefinitely), meaning that fewer new builds are required and the energy use per capita continues to decrease.
- With respect to industrial energy demand, we assume the nature and size of UK industry will roughly stay as it is. In other words, we do not assume that energy intensive industries (such as manufacturing) will play a greater or smaller role than they do today. UK industry in our scenario is simply a more energy efficient version of industry today.
- We assume the average person would like the UK to stay just as it is. This means that, as far as possible, we keep daily life very similar to now. There are, however, some things we simply cannot keep the same. We try though to make reasonable compromises, or choose options we think will have other benefits.

About the rest of the world in which our scenario exists:

- The rest of the world decarbonises alongside the UK, though we do not state how – it makes no difference if each nation or group of nations decarbonises alone, or as part of an international agreement.
- We assume that decarbonisation happens under a fair division of responsibility. This doesn't mean that everyone decarbonises at the same rate, but does mean that each nation keeps within its carbon budget (see *2.3 What does this mean for the UK?*). This means that GHG emissions associated with the production of goods that we import are accounted for globally, and the global carbon budget is still adhered to. A discussion related to the emissions associated with our imports can be found in *3.10.3 Carbon omissions*.
- The overall goal is a zero growth or steady state economy with a planned transition. Though we do not model global economics, we assume that the economy continuously becomes less energy (or carbon) intensive and that ultimately it is aiming to reach a steady state. Though this doesn't have much *explicit* impact on our scenario, we have known for a long time that economic growth, as it is currently generated, cannot continue while we live on a finite planet (Meadows, 1972).

About the transition to 2030:

- We do not explicitly model or make assumptions on *how we get there*. We do not assume a particular carbon price, emissions cap or suchlike. We create a scenario that technically achieves its aims – but it is not a road map of how to get there, which will likely depend on political persuasion and societal values. We can envisage a route that is either largely driven by market forces; by governmental regulation; by a voluntary large-scale change in aspiration by the UK population; or by widespread public demand for change in all sectors of society. We outline some of the options for policy frameworks in *3.8.2 Zero carbon policy*.
- We do, however, assume that the social and political priorities are different from those of today. We assume that over the course of the coming decades, the impacts of climate change will really start to bite, and that political and public motivation and action will become more aligned with what is physically necessary to rise

to the challenge of climate change. Every sector of society will have taken it seriously and will act accordingly. What is currently economically, socially or politically feasible takes second priority to what is physically necessary.

- We do not explicitly calculate the economic cost of our scenario or assume that there is a hard financial limit to our spending, though we do aim to avoid unnecessarily expensive solutions. Some technologies included are very expensive today because they are only used on a very limited scale. We assume that these will become a lot more financially viable when implemented on a large-scale. We assume that if the need is there, the market will follow. We also assume that the cost of *not* acting is unacceptably high.

Who made the rules?

These aims, rules and assumptions were laid out by the research team involved in the creation of this scenario. They broadly reflect the values held by the group, the social and environmental responsibilities we felt to be important, and some compromises and limitations that were necessary for the operation of the project. They are not meant to be a universal set of guidelines, or to reflect the only way of doing things. In fact, there was much discussion amongst the group and different viewpoints were held on various topics, even amongst what was assumed to be a set of fairly like-minded individuals. We might have chosen different constraints within which to construct a future scenario. Some of these are discussed in *3.10 Other scenarios*.

Measuring up today 3.2

Loz Flowers / Creative Commons ShareAlike 2.0 Generic

Before we figure out where we end up, we should take stock of where we are now – what the size of the challenge is for the UK, and what components might help us to get to net zero emissions, or hinder us.

So, what is the UK like today? In our scenario, we look at the UK in terms of three principal metrics:

- Our greenhouse gas emissions (GHGs).
- Our supply and demand of energy.
- Our use of land.

The first one depends heavily on the second two.

Figures 3.1-3.3 outline where we are today.

As stated above in *3.1 About our scenario*, in 2010, we emitted roughly 652 $MtCO_2e$ including those from international aviation and shipping (DECC, 2013). About 24 $MtCO_2e$ of carbon was also captured in the UK that year, yet the net effect on climate change was equivalent to 648 $MtCO_2e$.

We use roughly 1,750 TWh of energy every year, which requires a supply of about 2,530 TWh once losses in the system are taken into account (DECC, 2012a; DECC, 2012b). Our energy comes principally from fossil fuels: coal, oil and natural gas. This energy use creates 82% of our GHG emissions, and is comprised of energy use in households, businesses and industry, and transport.

These sectors, largely through industrial processes and the management of the waste we produce, also cause emissions that are not related to energy – about 7% of total annual emissions.

Just over 6% of our land is classified as 'urban' area, but as land is built on and grasslands and forests are cleared, more GHGs are emitted, contributing about 1% to our annual total emissions.

Over two-thirds of our land in the UK is dedicated to food production in some way, despite our importing about 42% of what we eat. Almost 70% of agricultural land in the UK is used to graze livestock (sheep and cows) for meat and dairy products. Even half of our cropland is used for livestock production – to grow feed. The agricultural use of land, and land use changes associated with it, contribute the largest portion of our GHG emissions after energy – roughly 10%.

Only 12% of our land is currently covered in forest, with about 90% of it harvested for timber. Just 8% of the UK's land is not managed or used productively in some way, which has significant implications for biodiversity and habitat protection – for example, over 80% of our peatland is damaged in some way due to our interventions, which further contributes to emissions. Forest (both harvested and unharvested) and some grassland are responsible for most of the carbon we currently capture in the UK.

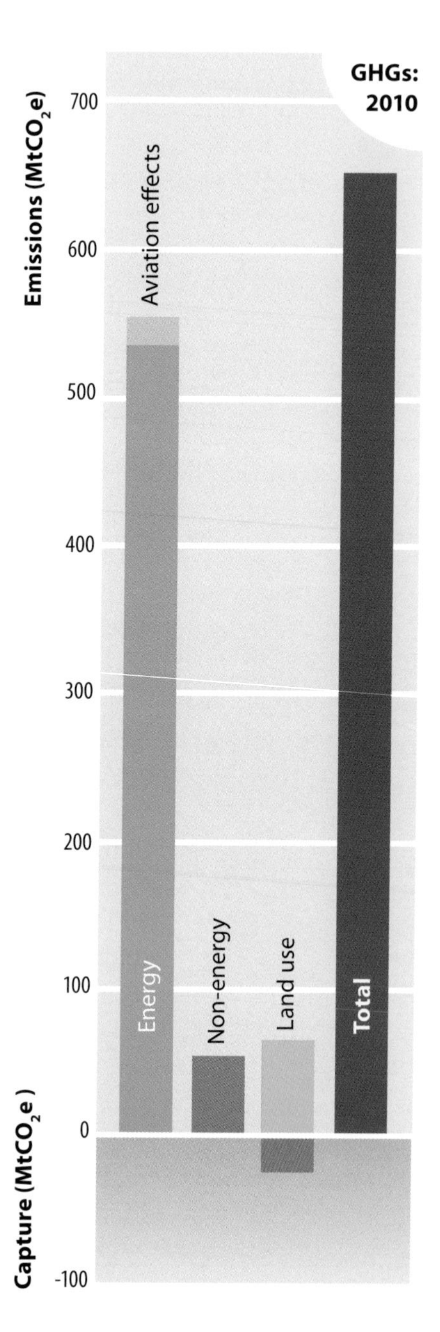

Figure 3.1: UK Greenhouse gas emissions in 2010, including international aviation and shipping, and the enhanced effect of emissions from aviation (DECC, 2013).

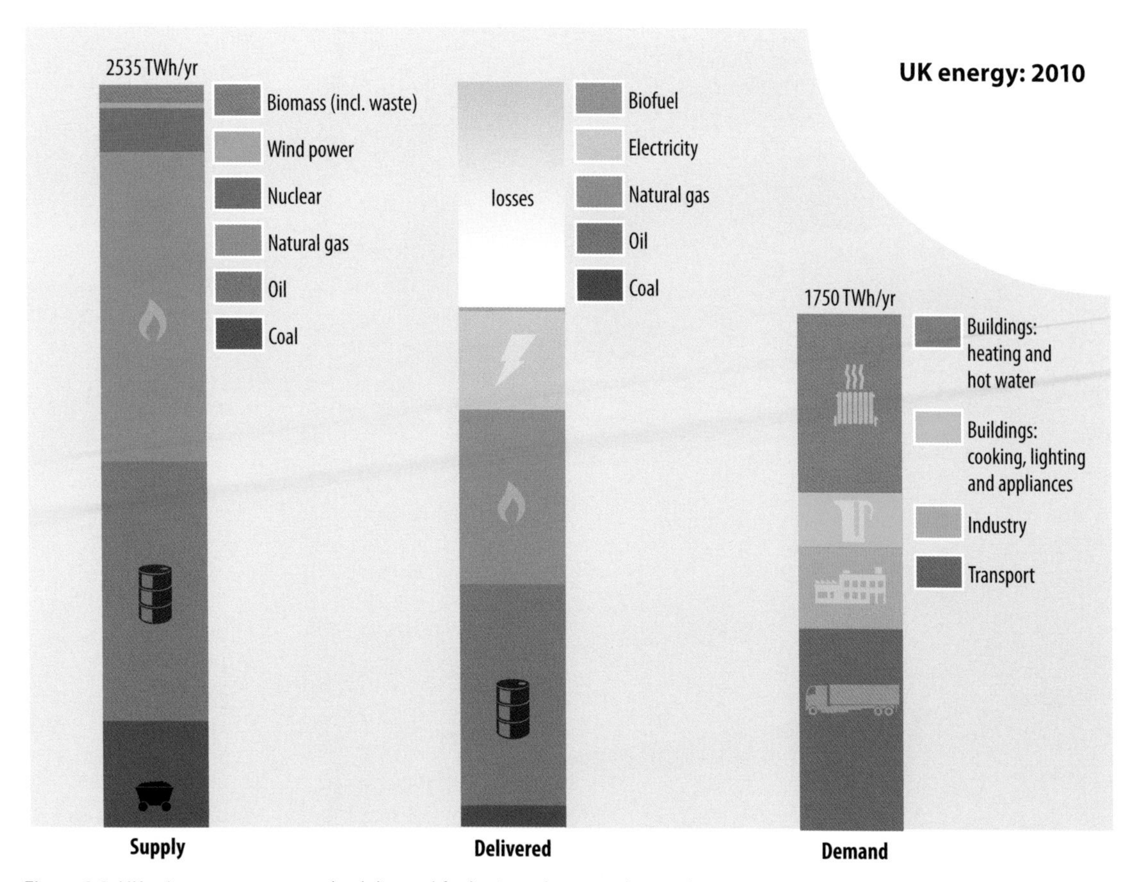

Figure 3.2: UK primary energy supply, delivered fuel mix and energy demand in 2010 (DECC, 2012a; DECC, 2012b).

Figure 3.3: Approximate land use today (not including water courses and coastal areas). Based on data from Morton et al. (2008), Forestry Commission (2007), DEFRA (2012), NERC (2008), Bain et al. (2011) and Read et al. (2009).

3.3 Power Down

Power Down is the reduction of our energy demand using efficient technology and making changes to the way we live. This is a vital part of the process of reducing GHG emissions from the energy system that powers our buildings, industry and transport. As outlined above, this energy demand – around 1,750 TWh in 2010 – accounts for roughly 82% of our current GHG emissions (DECC, 2013).

Power Down also makes it possible to fully meet our energy needs from renewable energy sources. As shown in sections *3.4 Power Up* and *3.6 Land use*, the UK could produce lots of electricity from renewables, such as wind power, but it has a limited amount of land available to grow biomass with which to make carbon neutral solid, liquid and gaseous fuels. The changes described in this section produce a 'fuel mix' that could be met by the UK's own renewable energy resources.

Power Down summary:

- Annual energy demand is reduced by about 60% from the current 1,750 TWh to around 665 TWh per year (see figure 3.4). An additional 105 TWh or so of ambient heat is used by heat pumps, making total energy use about 770 TWh per year (see figure 3.5 overleaf).
- A combination of efficient technology and behaviour changes can achieve large reductions in the energy used for heating and hot water, cooking, lighting and appliances, and transport.
- Industrial energy use is expected to remain similar to current levels – whilst industry will become more efficient, an increasing population and the need to build infrastructure will increase the demand for products.
- The 'fuel mix' resulting from Power Down means most energy is required as electricity (about 404 TWh per year), but some additional heat is required for buildings from geothermal and solar thermal generation – some 40 TWh every year.
- Buildings, industry and transport also require energy in solid, liquid and gaseous forms – 36 TWh of biomass for heat, 110 TWh of carbon neutral synthetic liquid fuel, 61 TWh of biogas or carbon neutral synthetic gas, and 14 TWh of hydrogen every year. Figure 3.5 shows this transition away from a fuel mix dominated by fossil fuels – oil and natural gas.

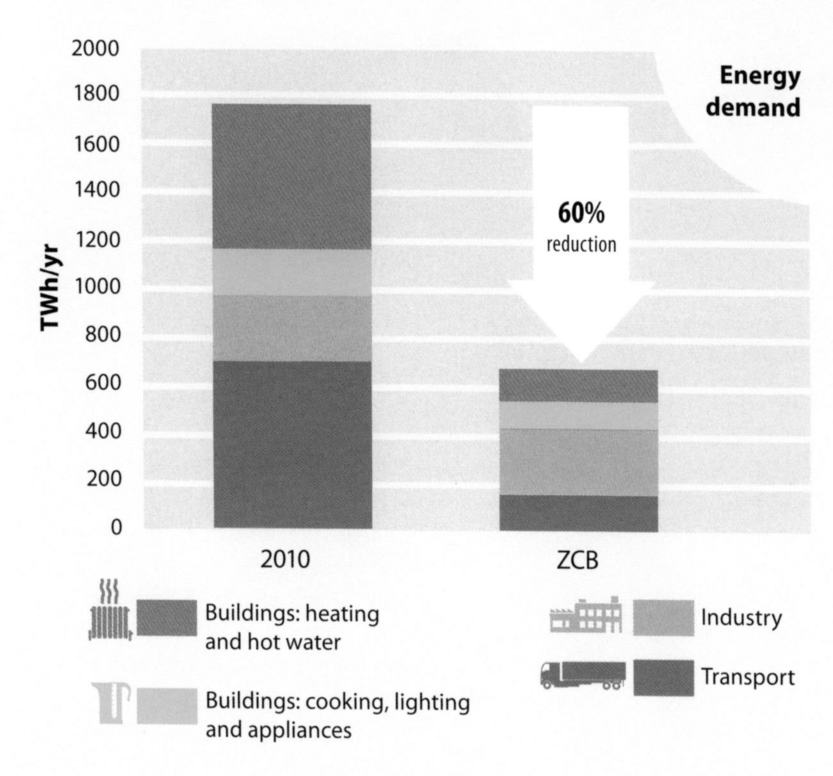

Figure 3.4: Total annual energy demand by sector in the UK in 2010 (DECC, 2012) and in our scenario.

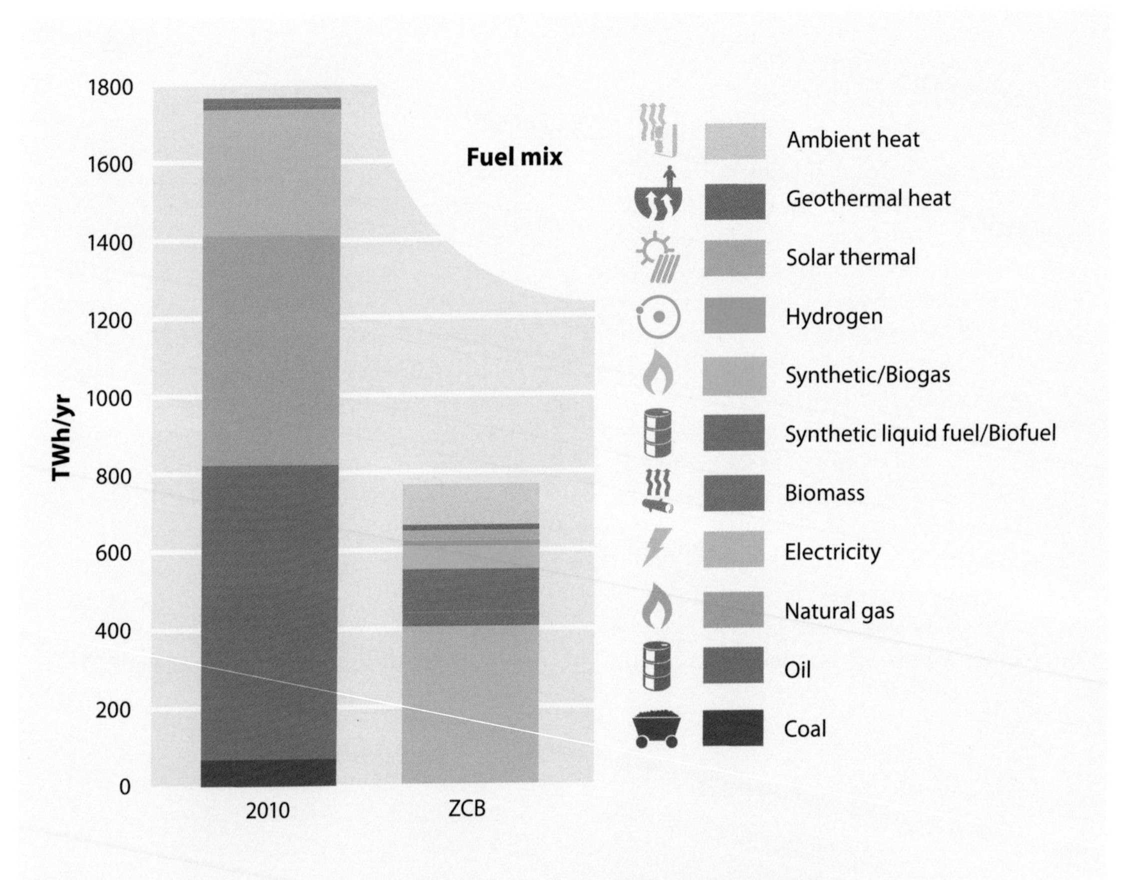

Figure 3.5: Annual energy use by fuel type in the UK in 2010 (DECC, 2012) and in our scenario.

3.3.1 Buildings and industry

This section covers energy demand and GHG emissions from the UK's building stock and industry. It describes how energy use in these sectors can be reduced and how the fuels used can change to allow the energy to come from renewable sources.

Summary

- Energy use in buildings and industry accounted for 61% of UK energy use and 54% of GHG emissions in 2010.
- High standards for new buildings and the retrofit of all existing buildings can reduce energy demand for heating by around 50%.
- Efficiency improvements in cooking, lighting and electrical appliances can significantly reduce their energy demand.
- Industry can also be made more efficient, but a growing population and the need to build infrastructure mean industrial energy demand is expected to be similar to today.
- In total, buildings and industry energy demand is reduced from around 1,050 TWh in 2010 to 510 TWh per year in our scenario (615 TWh including ambient heat).
- Most heating and hot water, all appliances, and most of industry will be powered by electricity (361 TWh per year), but we also require some biomass for heating buildings (about 10 TWh per year), and some heat from geothermal and solar thermal sources (40 TWh per year).

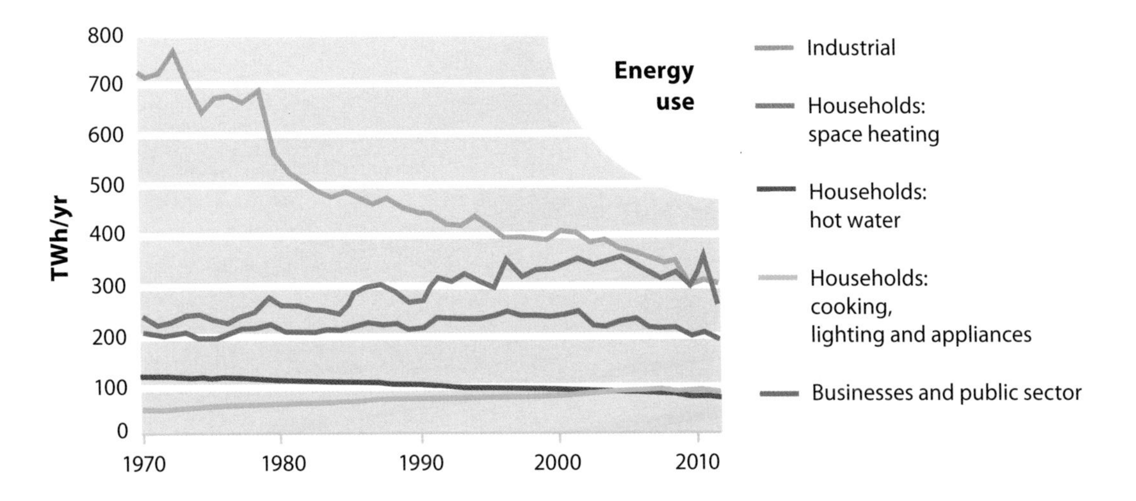

Figure 3.6: Annual energy use in UK buildings and industry over recent decades (DECC, 2012).

- Industry is also expected to need carbon neutral solid, liquid and gaseous fuels – 26 TWh, 12 TWh and 61TWh per year respectively.

What's the problem?

In 2010, 45% of the UK's energy use was in buildings – houses, offices, shops and public buildings (DECC, 2012). This energy was used for heating, cooling and ventilation, hot water, cooking, lighting and electrical appliances.

The UK currently has an aged and poorly insulated building stock. Small improvements have been made to reduce heat loss from buildings, but we have also tended to heat our buildings to higher temperatures. This means that energy demand for heating has risen over recent decades, although it may be starting to decline (see figure 3.6). Energy demand for hot water however, *has* declined over recent decades, thanks to more efficient hot water systems. Together, heating and hot water accounted for 34% of total UK energy demand in 2010 (ibid.).

Energy demand for cooking, lighting and electrical appliances was 11% of total UK energy use in 2010. This has decreased slightly over recent decades, whilst combined energy demand for lighting and appliances has increased slightly over the same period. The efficiency of cooking, lighting and appliances has improved, but we're also using more appliances – resulting in higher energy demand overall (ibid.). The figure of 11% also includes energy for cooling and ventilation, responsible for 0.5% of UK energy demand in 2010.

Industry is the only sector in which energy demand has reduced significantly in recent decades. This is a result of changes to the mix of products manufactured in the UK and improvements in how efficiently products are made. It should be stressed that total UK industrial output has increased slightly in recent decades, however the manufacture of some energy intensive products has decreased – for example, iron and steel production is now around a third of 1970 levels. Large efficiency improvements have also been achieved in many parts of industry. Together, these changes have reduced the overall energy demand of industry (ibid.).

Industry was still responsible for 16% of the UK's energy use in 2010 (ibid.), and it has remained a large source of the UK's GHG emissions – 20% in 2010 (ibid.). The emissions are produced by burning fuel for energy, but are also emitted directly by some industrial processes, such as cement production (see 3.5 *Non-energy emissions*).

Emissions from industry abroad

Although UK industrial output has only increased slightly over recent decades, our demand for products has increased significantly – we simply import more products from other countries. Emissions from manufacturing these products can be high, either because products are 'energy intensive' to make, or because they are made in countries that use high carbon energy sources.

These emissions are not included in UK 'production emissions', the figure most commonly used to represent total UK GHG emissions. Yet we still get the benefits of the goods we buy from abroad. Emissions associated with imported goods have risen around 60% since the early 1990s. When these emissions are included, total UK GHG emissions are shown to have *increased,* not *decreased* over the last two decades (DEFRA, 2010). *3.10.3 Carbon omissions* explores how these 'consumption emissions' can be accounted for and how reducing these emissions, as required to tackle climate change, might affect what we buy from abroad and what we make in the UK.

What's the solution?

To make a zero carbon Britain a reality we will need to reduce the energy demand from our buildings and industry, and put in place systems that allow us to meet this reduced energy demand with renewable energy and carbon neutral fuels.

Reducing heating demand

To reduce the energy demand for heating we must improve our building stock. By reducing the heat our buildings lose we will reduce the energy needed to keep them warm. Heat loss from buildings can be reduced by:

- Improving insulation.
- Reducing draughtiness.
- Recovering heat from air leaving the building through ventilation.

New buildings can have very low heat loss if they are constructed with excellent insulation and air-tightness, and are fitted with heat recovery ventilation. Passivhaus standard buildings, for example, have very low heating demand – around 10% of an average existing building today.

Heat loss from existing buildings must also be reduced, since the vast majority of today's buildings will still be in use in 2030, and beyond. Retrofitting existing buildings can include: cavity wall or solid wall insulation; floor and loft insulation; improved glazing (all of which reduce the 'fabric heat loss' of a building); and draughtproofing (which reduces the 'ventilation heat loss' of a building) – see figure 3.7. A programme to retrofit all dwellings with the above measures, as required, could reduce the average heat loss of the UK's housing stock by 50% (DECC, 2010).

Improved heating controls could also reduce energy demand by only heating rooms to the temperature required and when they are in use. Also, a more widespread culture of putting on a jumper rather than turning up the thermostat would have an impact on heating demand, and on our energy bills. With better heating controls and behavioural changes it could be possible to reduce average internal temperatures from the current average of 17.5°C to 16°C (ibid.). This would further reduce energy demand for heating. (Of course, reducing *average* internal temperatures will not stop us heating rooms to a higher temperature, such as 18-21°C, when we need to.)

Figure 3.7 shows how, in combination, the above measures can reduce space heating demand by around 50-60% per building on average.

Improved insulation can also help reduce the overheating of buildings in summer by keeping heat out rather than in. Adequate shading and ventilation are also needed though to prevent heat accumulating inside, and to allow the fabric of buildings to cool at night.

Zero carbon heating

Even with these changes we will still need to heat our buildings, and a bigger population will also lead to a slightly higher hot water demand (for washing,

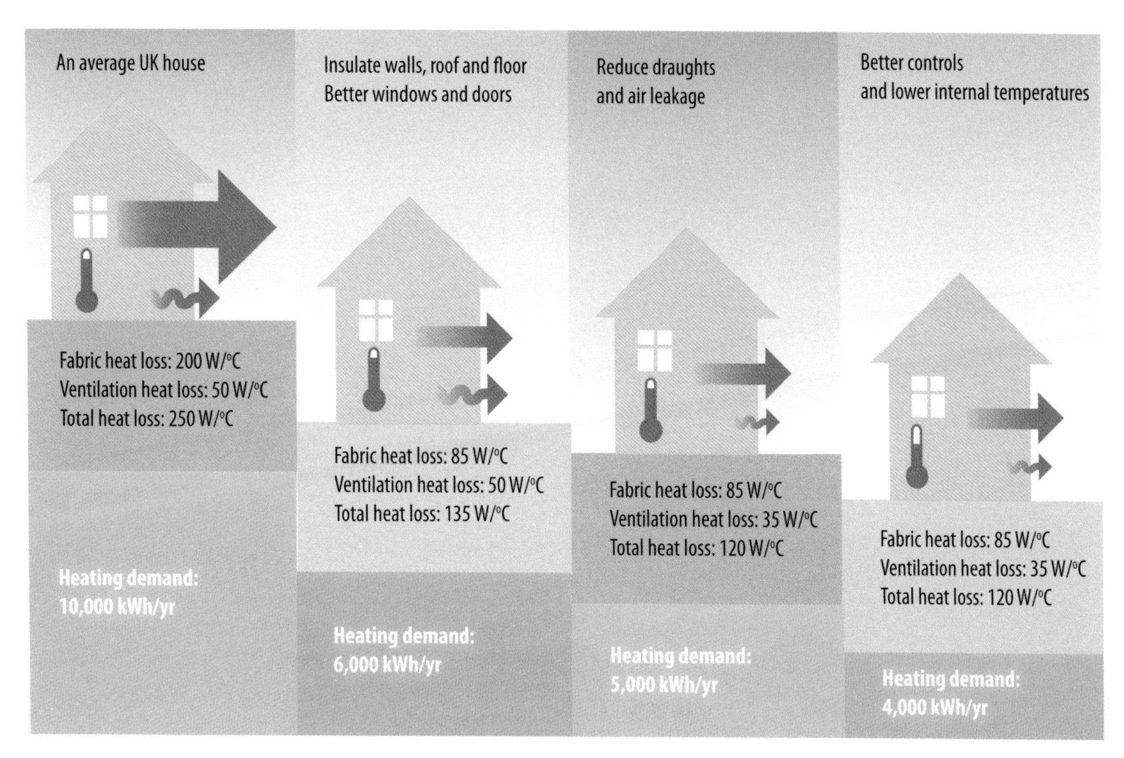

Figure 3.7: The impact of measures that reduce a building's heat loss and heating demand.

cleaning, etc.). This heating and hot water energy demand must be met without GHG emissions.

Solar heated hot water and geothermal heat can meet some of this demand, but most will be met by heat pumps. Heat pumps take ambient heat in air, water or the ground and 'concentrate' it, usually in water, to the required temperature. To do this they must use energy, but for each unit of electricity consumed heat pumps can typically deliver two to four units of heat – a very efficient way to generate heat from electricity. A mixture of biomass and direct electric heating systems can meet the remaining demand in situations where heat pumps are not practical, such as in buildings with large variations in energy demand, or which are not used regularly.

More efficient and smarter appliances

We can reduce the energy demand from lighting and electrical appliances significantly. Technological improvements can reduce 'in-use' power consumption, and better controls can minimise energy wasted by lights or appliances that are not being used. By maximising the currently available potential for efficiency, we can reduce lighting and appliance energy demand by around 60% per household, and by up to 30% in commercial and public sector buildings. Cooking can also be more efficient, using around 40% less energy per kitchen, and can be made fully electric. Systems used for cooling can be made around twice as efficient as today (ibid.).

As well as using more efficient appliances, it is possible to use smarter appliances. Such appliances are equipped with controls so that they can automatically reduce their energy demand to help balance the electricity grid. For example, at times of high demand, appliances such as fridges, freezers, washing machines and dishwashers would automatically use less energy over periods of just a few seconds or minutes and up to a few hours.

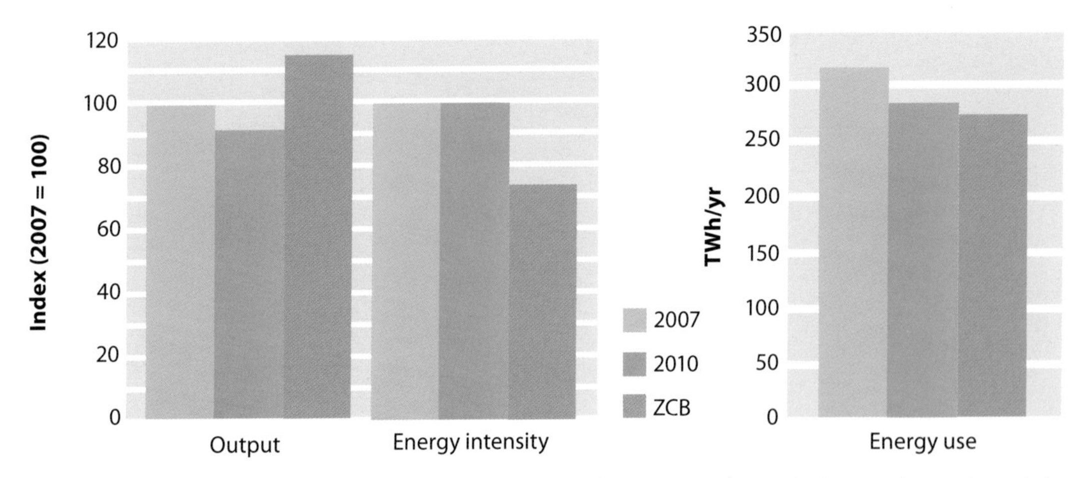

Figure 3.8: The amount of 'stuff' produced by UK industry (output), the energy used per unit of output (energy intensity), and the total UK industrial energy use for 2007 (representing pre-recession levels), 2010 (DECC, 2012) and in our scenario.

3.4.2 Balancing supply and demand further explains how this can help balance the supply and demand of electricity in a system incorporating lots of renewable energy.

Green industry

Energy use in industry depends on how much 'stuff' is produced (output), and on how much energy is needed to make each unit of 'stuff' (energy intensity) – see figure 3.8. These very much depend on:

- Changes in the demand for products.
- Shifts in what UK industry produces.
- Breakthroughs in efficiency.

An *overall* reduction in the demand for goods would decrease industrial output, and thus energy demand. Recycling, reusing and repairing items instead of throwing them away would likely lead to less demand for the production of goods. However, less demand for some goods can lead to more demand for others – known as the 'rebound effect'.

Changes to *what* UK industry produces could also reduce industrial energy demand. Reduced output of some energy intensive products, such as iron and steel, has already contributed to a reduction in emissions from UK industry. This trend could continue, for example if wood-based products were to substitute more conventional building materials (see *3.6.3 Capturing carbon*), further decreasing our 'production emissions'. Alternatively, to improve the economy and create jobs we might actually want to *increase* the manufacture of some energy intensive products. For example, we might wish to manufacture a high proportion of our renewable energy systems in the UK.

Improvements in energy efficiency would mean the same amount of 'stuff' could be produced but using less energy – therefore reducing energy intensity. The UK has already reduced energy intensity in most industrial sectors over recent decades, and further energy intensity reductions of up to 25% are considered ambitious but feasible (ibid.).

Switching the type of energy we use

All energy used for heating, hot water, cooking, lighting and appliances could be supplied as electricity, rather than from sources like gas boilers or gas cooking hobs. *3.4 Power Up* describes how this electricity demand could be met by UK renewable energy sources.

In industry, however, the form in which energy is supplied can be important. Whilst it is possible to increase the proportion of industry powered by electricity, it may not be possible to fully electrify all industrial processes. For example, it may not be practical to use electricity for some high temperature

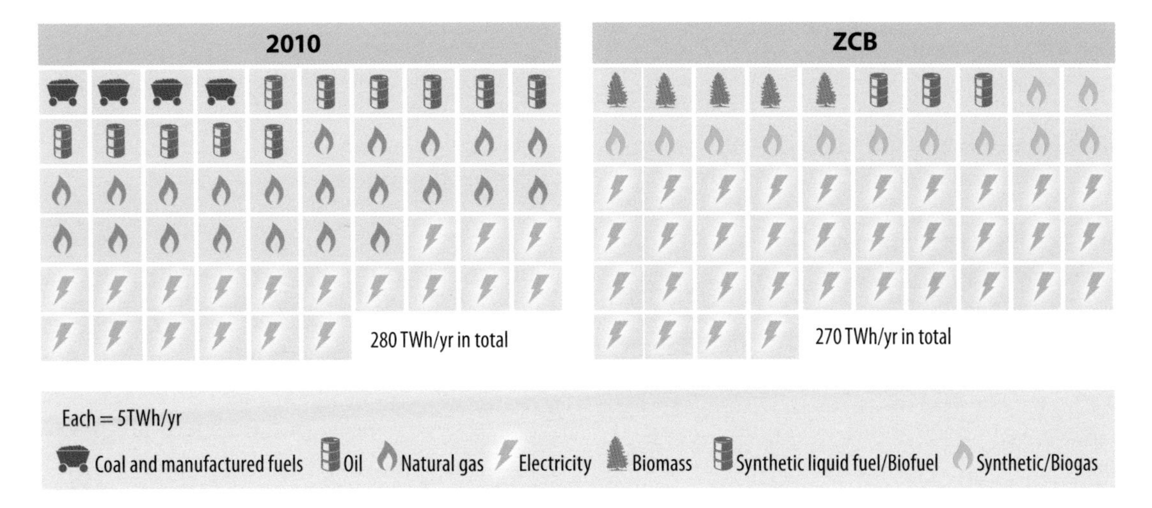

Figure 3.9: Mix of fuel used annually by UK industry in 2010 (DECC, 2012) and in our scenario.

What's 'carbon neutral'?

Synthetic gas, biogas, synthetic liquid fuel and biofuel can all be 'carbon neutral'. The CO_2 emitted by burning them was initially taken in by the biomass as it grew, and the electricity used to produce the hydrogen required (via electrolysis) can be renewably generated (see *3.4.2 Balancing supply and demand* and *3.4.3 Transport and industrial fuels*). Over the long-term there is no net increase in greenhouse gases in the atmosphere. The advantage of synthetic gas and liquid fuels over pure biogas and biofuel is that less biomass is required to produce the same amount of gaseous or liquid fuel.

processes, such as some iron, steel and ceramics manufacture.

The production of certain chemicals also uses fuels like natural gas (methane) as a feedstock. It is estimated that roughly 50 TWh of the annual industrial fuel demand could be most feasibly met using carbon neutral synthetic gas (made by combining biomass and hydrogen – see *3.4.2 Balancing supply and demand*) or biogas made purely from biomass (NERA, 2010). Solid biomass can also efficiently provide heat for industry using Combined Heat and Power (CHP) systems, which also generate electricity. In addition, some industrial machinery may require liquid fuels equivalent to oil. Other 'high electrification' scenarios for industry also suggest some biomass, synthetic gas or biogas and synthetic liquid fuel (made in a similar manner as synthetic gas, combining biomass and hydrogen) or biofuel will be needed to meet future industrial energy demand (DECC, 2010).

Building in flexible demand

3.4 Power Up shows how renewables can meet the energy demand in our scenario. To make this easier, electricity demand should be made as flexible as possible, so that it can move up or down in response to the availability of electricity from renewables.

As discussed, smart appliances can help balance the grid by responding to signals and changing the times at which they draw power. The electricity demand for heating and hot water can also be made more flexible by having large heat stores (usually tanks of hot water), so that heat can be produced and stored at times when the electricity supply from renewables is high. Such heat stores could be inside buildings, or buildings could be connected to external heat stores supplying anything from a few houses to whole districts. Buildings themselves can also act as leaky but useful heat stores. If they only lose heat slowly, buildings can be heated when electricity is plentiful rather than exactly when heating timers are set.

Industrial electricity demand can also be made more flexible, with production decreasing at some times and increasing at others. This would help balance electricity supply and demand. This already occurs today but it could have a bigger role in the future, with industry adjusting its electricity demand both up and down, and more often.

The roles that heat stores and flexible energy demand from industry can play in a renewable energy system are discussed further in *3.4.2 Balancing supply and demand*.

Our scenario

In our scenario, energy demand for heating buildings is reduced by around 50% because:

- All new houses will be built to Passivhaus standard, or similar.
- A mass retrofit of all existing buildings (including offices, schools, etc.) will take place.
- Better heating controls and changes to behaviour will reduce average internal temperatures.

The trend in improved efficiency of hot water production continues, but a bigger population will lead to a slightly higher hot water demand – a 16% increase. Heating and hot water energy demand is about 235 TWh per year in total, though this will vary from year to year depending on outside temperatures. Heat pumps meet the majority of this demand (using about 50 TWh per year of electricity and 105 TWh per year of ambient heat); direct electric heating requires around 30 TWh per year of electricity; 10 TWh is provided directly with biomass; and the remaining heating demand (40

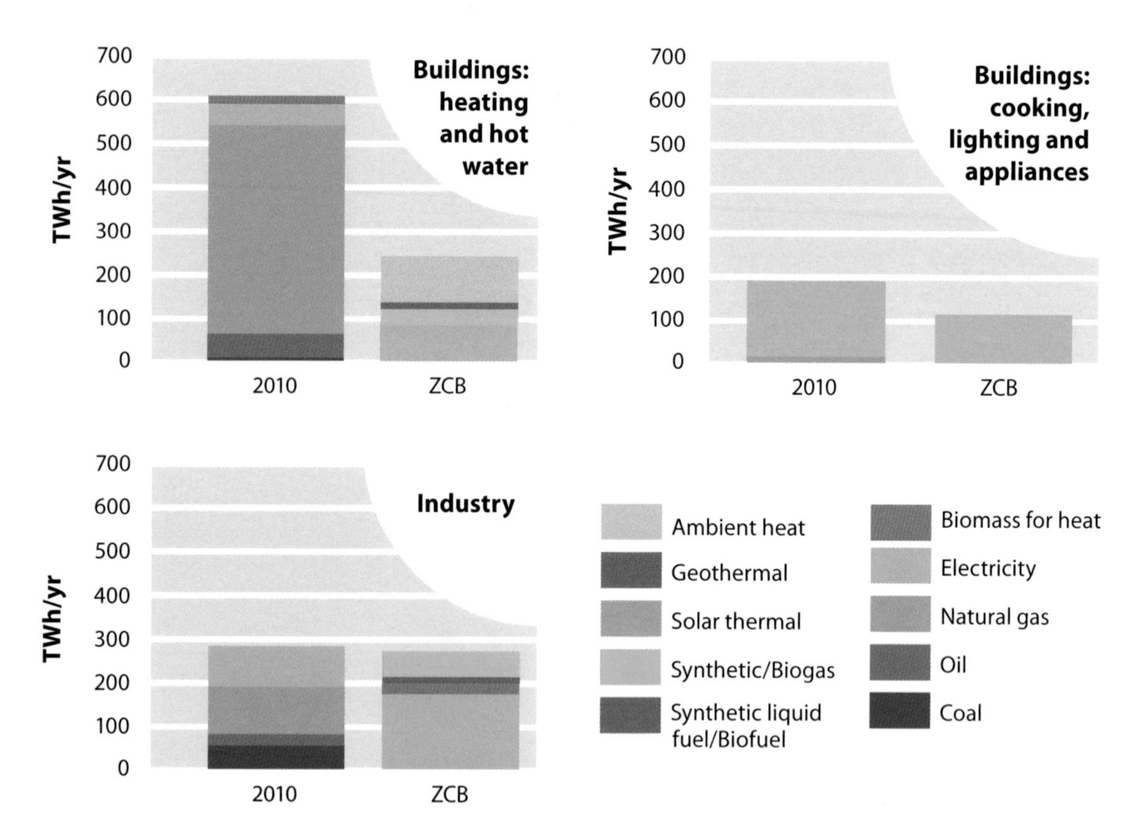

Figure 3.10: The change in energy demand for heating and hot water; cooking, lighting and appliances; and industry between 2010 (DECC, 2012) and our scenario: by amount and type of fuel.

TWh per year) is met by solar thermal and geothermal heating (see figure 3.10).

Despite potentially higher average temperatures due to climate change, better building insulation and well-designed shading and ventilation means that cooling demand remains at current levels. Since cooling systems become more efficient, energy demand for cooling and ventilation falls from 9 to 5 TWh per year.

Efficiency improvements reduce energy demand for cooking, lighting and electrical appliances by around 40% to 105 TWh per year.

We assume that UK industrial output per person returns to 2007 (that is, pre-recession) levels, although population growth means total output is 16% higher than in 2007. Exactly *what* will be produced is uncertain and may be very different from that produced in 2007. For example, the manufacture of renewable energy systems increases, but demand for other goods may also decrease if we replace some more conventional building materials with wood products (see 3.6.3 *Capturing carbon*), if our society becomes less 'consumerist', and if we place greater emphasis on the longevity and reparability of products – recycling and reusing more. In our scenario, a strong push for further efficiency is assumed to reduce industrial energy intensity by 25% on average. Total industrial energy demand is 270 TWh per year (see figure 3.8).

Figure 3.10 shows the changes in energy use between 2010 and in our scenario, as well as the change in fuel type. Electricity supplies the majority of energy for heating, hot water, cooking, lighting and appliances. The proportion of industrial energy demand met by electricity also increases to 63% – to about 171 TWh per year (see figure 3.9). Other industrial processes require:

- Around 61 TWh of biogas or synthetic gas per year.
- Smaller amounts of biomass for heat (26 TWh) and synthetic liquid fuel (12 TWh) every year.

Energy demand is much more flexible than it is today with smart appliances, large heat stores inside or connected to buildings, and more flexible industrial electricity demand.

Overall, energy use in buildings and industry is reduced by about 50% to around 510 TWh per year (or 615 TWh per year if the ambient heat used by heat pumps is included) – 361 TWh (about 70%) of which is electricity demand.

3.3.2 Transport

This section covers energy use and emissions in the UK from transport, including international aviation and shipping. It describes how changes to our transport system can reduce energy demand and allow the energy to come from renewable sources.

Summary

- In 2010, 39% of UK energy demand and 25% of UK GHG emissions were from transport. Surface passenger transport accounted for about 55% of transport energy demand, aviation about 20%, and freight around 25%.
- Increased walking, cycling, and use of public transport can reduce our energy demand and GHG emissions, as well as making our urban environments more pleasant and making us healthier.
- Most transport can be switched to very efficient electric vehicles. Hydrogen powered vehicles may also have a small role to play, but some road vehicles, as well as ships and aeroplanes, will continue to need liquid fuels.
- International aviation can be made more efficient, but its need for synthetic liquid fuel (which requires biomass), as well as additional climate impacts of GHGs emitted high in the atmosphere, mean it must be reduced to around a third of current levels.
- In our scenario, the need for freight transport is reduced as all of our energy and more of our food comes from the UK. Freight transport vehicles also become more efficient and 20% of road freight switches to rail.

- In total, UK domestic and international transport energy demand is reduced from around 700 TWh in 2010 to 155 TWh per year in our scenario – a 78% reduction. 43 TWh of electricity is required per year, and energy demands in the form of synthetic liquid fuel and hydrogen are 98 TWh and 14 TWh per year respectively.

What's the problem?

In 2010, 39% of UK energy demand and 25% of UK GHG emissions were from transport (DECC, 2012; DECC, 2013). Energy for transport is overwhelmingly derived from petroleum products such as petrol, diesel and kerosene. Fuel use and GHG emissions from transport have increased over recent decades, peaking in 2007 and declining a little since. Road vehicles use most of this fuel (70%) but trains, boats and planes also use significant amounts (DfT, 2012) (see figure 3.11).

On average, a British person travels around 6,500 miles a year by car or van. This figure has declined slightly in recent years after increasing for many decades (ibid.). Efficiency improvements mean that, on average, the UK's cars are slowly using less energy and emitting less CO_2 per mile travelled – average CO_2 per mile is decreasing at about 1% a year (CCC, 2012). Nevertheless, cars and vans account for about 50% of all transport energy use (DfT, 2012).

Around a further 1,400 miles is travelled on average per person per year by foot (200 miles), bicycle (50 miles), motorbike (50 miles), bus (450 miles), and train (650 miles). The amount of travel by these modes has been broadly stable over recent decades, except for train travel, which has nearly doubled. Taken together, these forms of transport accounted for just 5% of all transport energy use in 2010. Interestingly, in the 1950s and '60s, travel by bicycle and bus were at much higher levels than today (ibid.).

In recent decades, people have also been flying more – passenger numbers peaked in the mid-2000s and have declined a little since. The number of passenger arrivals and departures at UK airports has more than doubled since 1990 to around 200 million in 2010. Around 20% of flights are domestic and 80% international. Aviation accounts for about 20% of all transport energy use (ibid.). CO_2 emissions from burning aviation fuel contributed 5% to the UK's GHG emissions in 2010 (DECC, 2013). However, other factors, such as aircraft contrails and emissions high in the atmosphere inducing cirrus clouds, may multiply aviation's impact on climate change by as much as two (Lee, 2010). In 2010, this would have had an additional impact equivalent to almost 20 $MtCO_2e$. Aviation also brings high noise pollution.

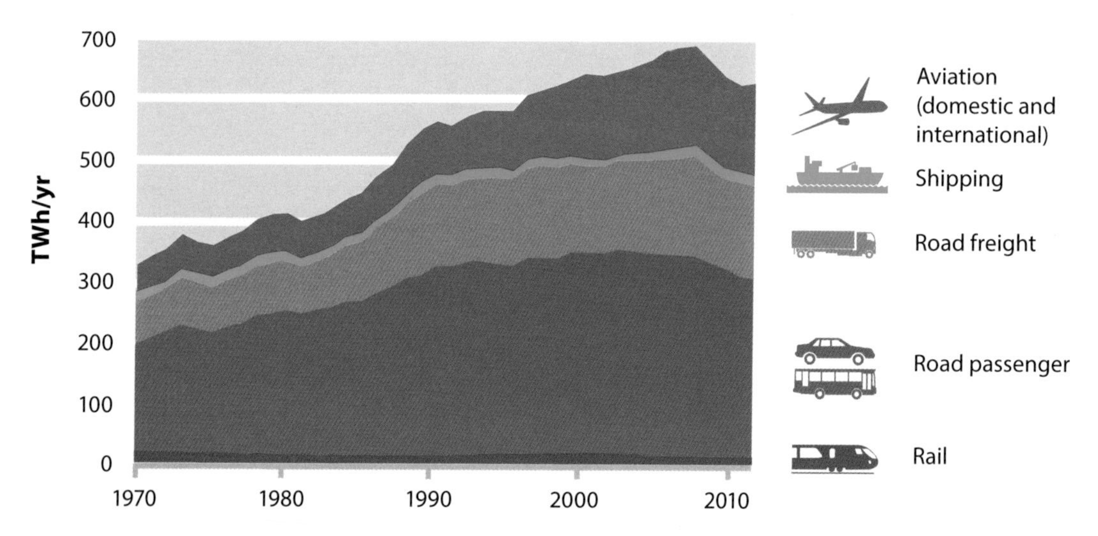

Figure 3.11: Energy demand for UK transport over recent decades (excludes international shipping (DECC, 2012)).

Although fuel use by freight (Heavy Goods Vehicles (HGVs), trains, ships and planes transporting goods) has decreased in the last few years due to the recession, in general it has been increasing over recent decades (DECC, 2012). The amount of goods transported, and the distance they are moved, follow a similar trend (DfT, 2012). In total, freight accounts for around 25% of our transport energy use (DECC, 2012).

What's the solution?

Radical changes will be needed to the amount and the way we travel and move goods in a zero carbon Britain. This is essential to reduce energy demand. Changes are also needed to our transport system to make our urban environments more pleasant places to live and work, and to help us be more active and healthier.

How we travel, and how much

Improved communication technology (video conferencing, Skype, etc.) can make some journeys unnecessary. Living closer to where we work and play would also lead to less travel – and less time spent commuting.

Better infrastructure in towns and cities (cycle lanes and pedestrian areas, for example) can encourage people to walk and cycle shorter journeys. This has health as well as environmental benefits, and would decrease noise pollution in urban areas. Better public transport – bus, coach and rail – can also get people out of their cars, reducing road congestion, energy use and GHG emissions.

When we do use cars, we can make better use of them by increasing the number of occupants. By arranging car sharing, either informally or via car share schemes, the average occupancy of cars could improve from the current average of 1.6 people per vehicle (Dft, 2009a).

These changes will reduce energy demand from transport. Just as importantly, they could create more pleasant places in which to live, and could make us healthier.

Alternative fuels – biofuels, synthetic liquid fuels, hydrogen or electric vehicles?

Even if we travel less and more efficiently, we will still need lots of energy for transport. One possibility to reduce GHG emissions from transport is to use biofuels – fuels made from plant material ('biomass'), or carbon neutral synthetic liquid fuels – fuels made from biomass combined with hydrogen (see *3.4.3 Transport and industry fuels*).

European Union targets have seen biofuels mixed into the supply of petrol and diesel to around 3% of the mix (Dft, 2012). Even at this low level, serious concerns have been raised about the effects of biofuel production on land use, and consequently food prices and biodiversity. It is clear that fuels made using biomass cannot replace all the petrol and diesel we use in cars, let alone meet all of our current transport energy needs. This would remain the case even after the changes to travel described above, and even if petrol or diesel vehicles were made more efficient. It is, therefore, necessary to change the *type* of fuel our vehicles use.

Electric cars and buses offer a solution. They are around three times as efficient as cars and buses that run on petrol and diesel – and they could be up to six or seven times as efficient as the average vehicle on the roads today (DECC, 2010). In addition, their batteries can be charged with electricity from renewables.

The distance that electric road vehicles can travel before they need recharging makes longer journeys more difficult. However, statistics show that around 90% of journeys made by cars are less than 100 miles long (DfT, 2009a). Modest improvements in battery technology would mean all such journeys would be achievable on one charge (current electric cars can travel about 80 miles per charge). The scheduled recharging of buses and coaches would be straightforward, if planned into timetables. The development of an adequate electric vehicle charging infrastructure would be essential, but this poses no technical challenges.

For some specialist vehicles, such as those used off-road, heavy commercial vehicles (such as HGVs, tractors and diggers) or those requiring longer range, hybrid 'biofuel-electric', fully biofuel or synthetic liquid fuel powered vehicles could be used.

That said, a further reduction on even this small reliance on fuels derived from biomass is desirable and hydrogen fuel cell vehicles are a possibility. However, widespread use of hydrogen vehicles requires an entirely new infrastructure to distribute the hydrogen fuel. Since electric vehicles are a more efficient and simpler option for mass transport, a widespread hydrogen distribution network is unlikely to be developed. Therefore, hydrogen vehicles are likely to be used only on a small-scale.

Less aviation

No serious alternative to liquid fuel currently exists for planes. This is because aviation fuel must have a very high energy density by weight and volume – it must be small and light. A plane carrying enough electric batteries to power its journey would be too heavy to fly, whilst a plane carrying enough hydrogen would be too large to fly at speed – although it could fly slowly, like an airship.

Aviation can use less energy if we improve aircraft efficiency and manage flights better. This could improve aviation's fuel use per passenger by around 1% per year for the next few decades (DfT, 2009b). However, efficiency improvements will run into the physical limits that determine the energy required for flight. Therefore, to drastically reduce this sector's GHG emissions we must replace current petroleum-derived liquid fuel with sustainable biofuel from biomass, or synthetic liquid fuel made from biomass and hydrogen. As shown in *3.6 Land use* there are constraints on the land available to grow the biomass required for these fuels.

Another concern is that even a complete switch to biofuels or synthetic liquid fuels does not stop the additional impact on climate change of contrails or gases emitted high in the atmosphere. Therefore, the only way to reduce the climatic impact from aviation is to fly less.

Rail or coach can replace journeys within the UK currently made by plane, as well as relatively local international flights. Eurostar connections provide an example of how European journeys currently

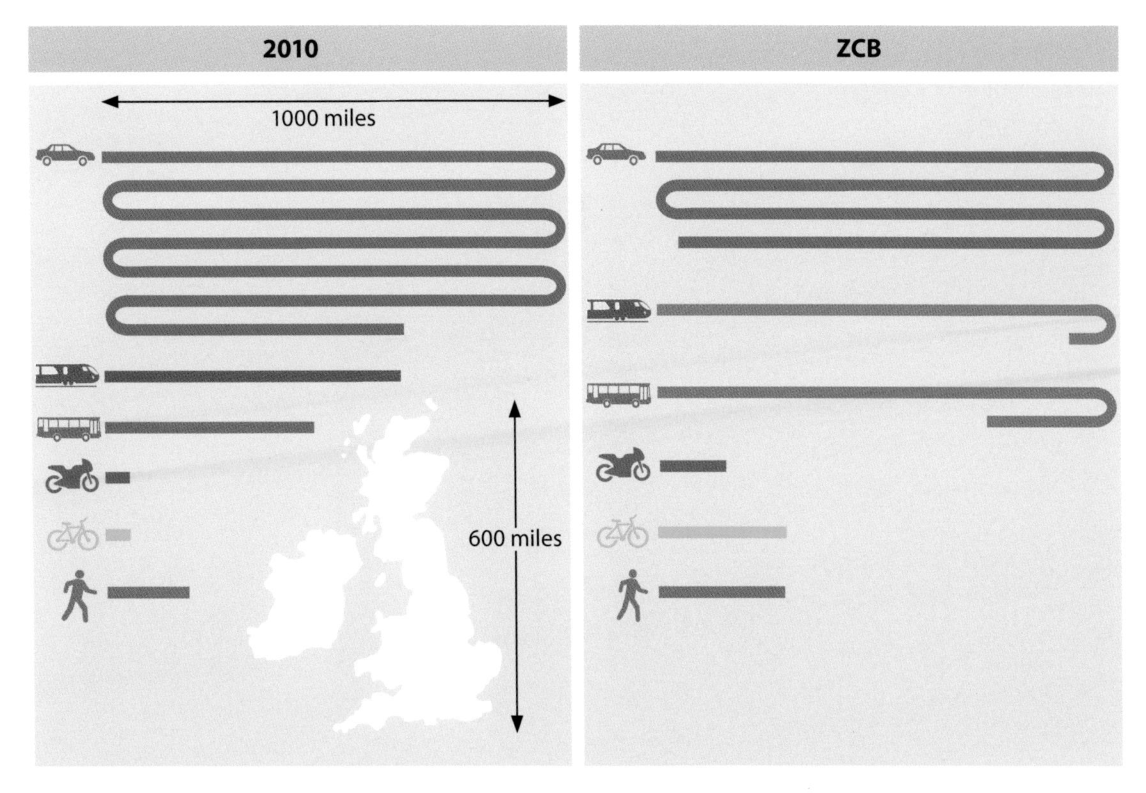

Figure 3.12: Average distance travelled per person per year by various modes of transport in 2010 (DfT, 2012) and our scenario.

made by plane could be made by high-speed rail instead. Nevertheless, a reduction in flying does challenge the strong social norm and perceived *right to fly* that has developed over recent decades.

Changing how we move 'stuff'

To reduce energy demand from transporting goods (freight) we can:

- Reduce the amount of goods we move.
- Reduce how far goods travel by sourcing them closer to home.
- Improve the efficiency of vehicles used.

Whilst Heavy Goods Vehicles (HGVs) will become more efficient in the future – with more efficient engines, better aerodynamics and more efficient operation – they will still run mainly on liquid fuels. Given the limits on biomass for biofuel and synthetic liquid fuel supply, and to reduce road congestion, shifting more freight to railways makes sense. Increasing rail freight by 200% over 2010 levels is considered feasible (DECC, 2010).

Due to limitations on aviation, moving freight by air should be eliminated for all but essential items.

Ships are likely to continue to require similar liquid fuels – in the form of biofuel or synthetic liquid fuel. Fuel use in shipping can be reduced through more efficient engines and better management to reduce ships travelling only partially full or empty. Changes to the demand for some goods can also reduce the need for shipping.

Our scenario

In our scenario, the distance travelled per person decreases by around 15% from 2010 levels, as better communication tools reduce the need for some journeys and people live closer to where they work and socialise. People walk and cycle more, and the use of public transport – buses, coaches and rail –

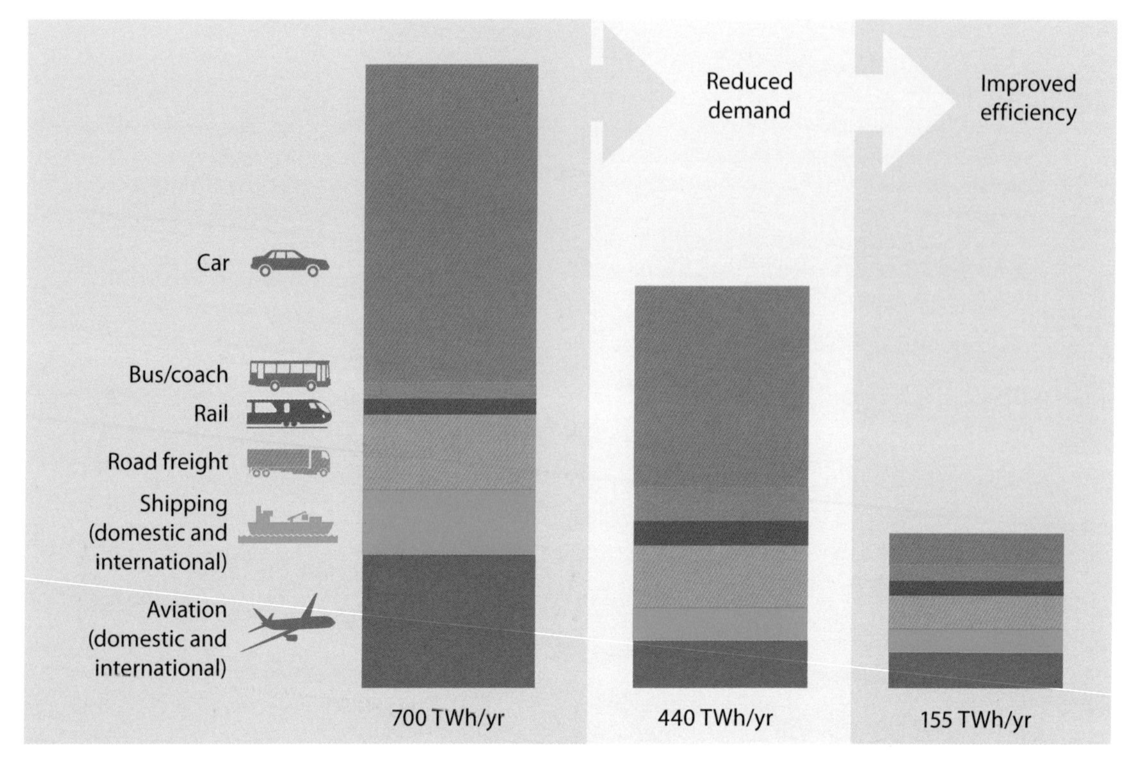

Figure 3.13: Reduction in energy demand for personal and commercial (freight) transport in our scenario (with initial figures from DECC, 2012).

increases from 14% to 33% of domestic travel. As a result of these changes, car travel is reduced from 82% to 56%. In addition, the average occupancy of cars increases from 1.6 to 2 people per vehicle. See figure 3.12 for a summary of these changes.

Around 90% of road passenger transport is electric vehicles – cars, vans, coaches and buses. The rail network is also close to fully electrified (95%).

Hydrogen powered vehicles are favoured to reduce the demand on land for biomass but, since a full infrastructure for hydrogen distribution is not envisaged, some synthetic liquid fuel powered vehicles are used. Carbon neutral synthetic liquid fuels and hydrogen power the remaining road passenger vehicles, such as those requiring longer range and heavy commercial vehicles.

Our scenario includes no domestic aviation – most of the journeys currently made by domestic flights are now made by rail. The number of miles flown for international aviation falls by two-thirds. In combination with efficiency improvements, this reduces aviation liquid fuel demand by around 75%.

Additional impact of flying in our scenario

Carbon neutral synthetic fuel is used to fuel planes in our scenario. It is 'carbon neutral' because the CO_2 emitted by burning it was initially taken in by the biomass as it grew, and the hydrogen used in its manufacture was produced using renewable electricity. Over the long-term, this means there is no net increase in GHG emissions in the atmosphere. However, contrails or gases emitted high in the atmosphere by flying may lead to an additional impact on climate change (Lee, 2010). Even with substantial reductions in flying in our scenario, there is a remaining impact that is equivalent to about 7.2 $MtCO_2e$.

In our scenario, changes to our energy and food system eliminate the need to move some goods (such as fossil fuels), but increase the need to move other goods (such as biomass). In general, since all of our energy and more of our food comes from the UK, we need less freight transport.

Total road and rail freight reduces marginally (about 5%), but rail freight more than doubles as around 20% of road freight switches to rail. HGVs and other heavy commercial vehicles (tractors and diggers, for example) are mostly powered by carbon neutral synthetic liquid fuel (80%), with some hydrogen powered vehicles (20%). Freight moved by air is all but eliminated, and changes to the type of goods that need moving means shipped freight decreases by over 50%. Ships are powered by synthetic liquid fuels.

Overall, energy demand from transport falls by 78% from 2010 levels, to 155 TWh per year (see figure 3.13). With much of the transport system electrified, transport electricity demand rises to 43 TWh per year. Energy demand for synthetic liquid fuel is 98 TWh per year (59 TWh for heavy commercial vehicles, and 39 TWh for planes) and for hydrogen is 14 TWh per year. Figure 3.14 summarises this change.

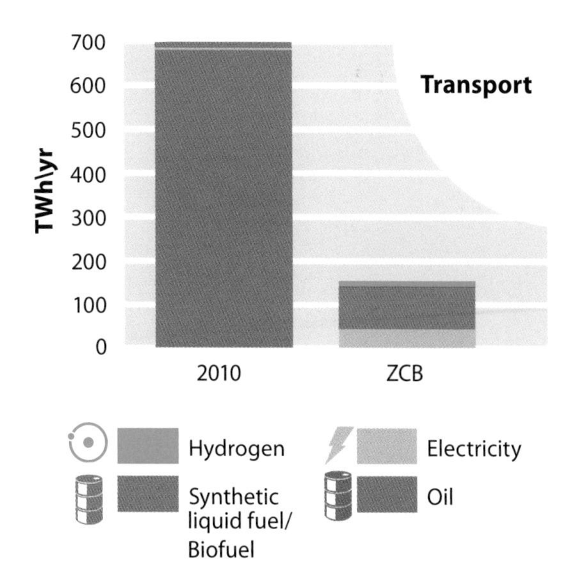

Figure 3.14: Change in total energy demand for transport and the types of fuel required in 2010 (DECC, 2012) and our scenario.

3.4 Power Up

The preceding section, *3.3 Power Down* outlined how we can reduce our energy demand (what we use), and this means we can reduce the amount of energy we produce, and thus the amount of greenhouse gases (GHGs) we emit. However, it is important not to underestimate how much energy is still required. In our scenario, the final energy demand – the amount of energy we use, including ambient heat, but excluding all exports and losses – is around 770 TWh per year. This is less than half of today's final energy demand, which is around 1,750 TWh per year (DECC, 2012). However, it is still a very large amount of energy compared to, for example, the amount of energy produced by wind turbines in the UK today (around 10 TWh in 2010) or even globally (342 TWh in 2010).

This Power Up section outlines how renewable energy sources can meet 100% of this energy demand, reducing the GHG emissions from our energy production to zero.

In our scenario, the largest contribution will come from offshore wind turbines, which can produce around half of the energy we need. Figure 3.15 shows our final energy mix. However, this reliance on renewable energy from variable sources, like wind power, makes it challenging to ensure that energy supply always meets demand. A range of demand management methods and energy storage technologies play a role in solving this problem.

Biogas from biomass, and chemical processes for creating carbon neutral synthetic gas and carbon neutral synthetic liquid fuels from biomass and hydrogen (produced using surplus renewable electricity), allow us to balance energy flows and replace fossil fuels in systems that are difficult to electrify. Although there are significant losses in these processes, without them we would not be able to meet all demands.

Power Up summary:

Today, almost 90% of our energy comes from fossil fuels. Together, Power Down and Power Up eliminate all emissions from our energy system. In our scenario:

- Renewable energy provides a primary energy supply before conversion losses of around 1,160 TWh per year, allowing us to meet 100% of a final energy demand of 770 TWh per year (665 TWh per year, not including ambient heat for use in heat pumps) using only 'zero carbon' energy sources.
- Wind energy plays a central role, providing around half of the primary energy supply (581 TWh per year). The rest is generated using various renewable sources of energy. Figure 3.15 shows the change in energy mix between 2010 and in our scenario.
- Matching supply and demand in our scenario with a large share of energy from variable sources is technically challenging, but possible, incorporating chemical processes that create synthetic gas from biomass and hydrogen as back up. Only 27 TWh per year is required of this, but it plays a critical role when demand is high and supply from renewables is low (for example, when it is cold but not windy).
- Most of the energy in our scenario is produced in the form of electricity – about 60%, but there is also a significant amount of energy supplied in other forms – biogas is produced from biomass, and synthetic gas and liquid fuels are produced from surplus electricity and biomass. There are losses in the conversion processes, but demands from industry, transport and energy back up require these specific fuel types.

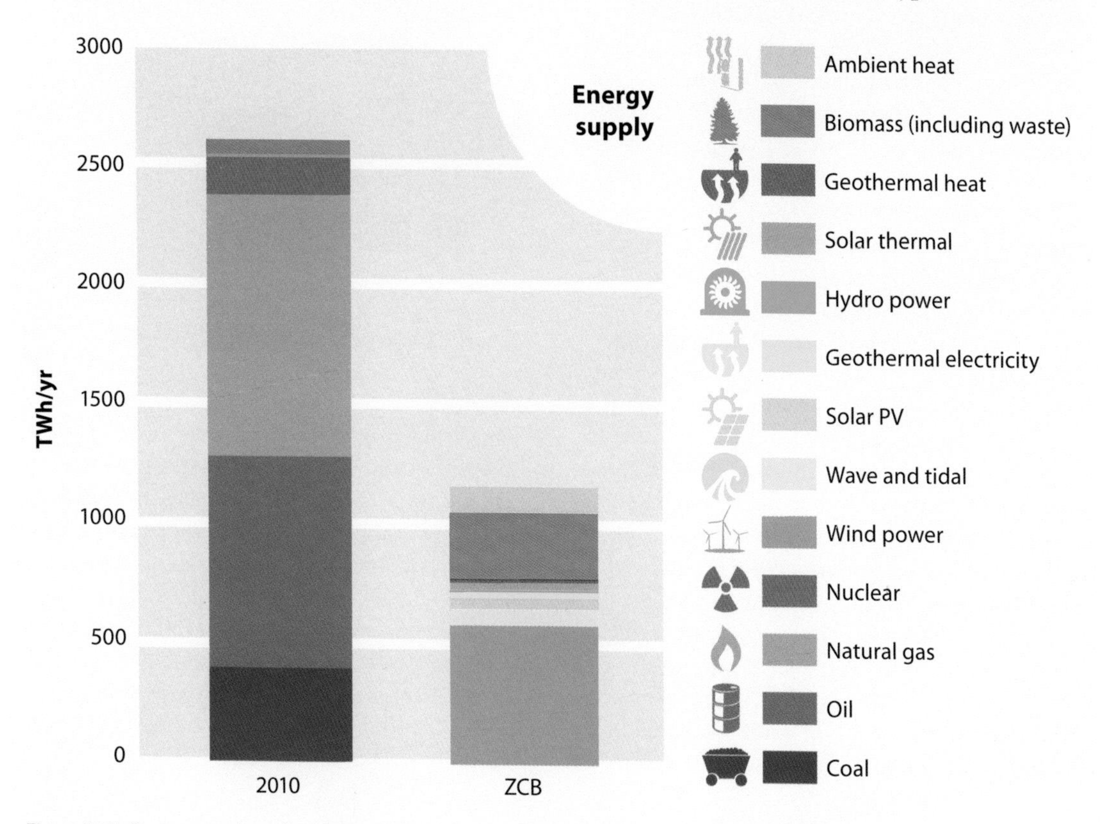

Figure 3.15: Energy supply today (DECC, 2012) and in our scenario.

3.4.1 Renewable energy supply

The section describes how we can Power Up the UK using renewable energy generation – providing all our energy supply from zero carbon technologies.

Summary

- Renewable energy only contributes a small proportion to our total energy supply today.
- In our scenario, we produce about 1,160 TWh per year from renewable sources to meet 100% of the 770 TWh annual energy demand.
- Today, only 20% of our energy is in the form of electricity, but in our scenario most energy (738 TWh per year) is produced in the form of electricity, generated by a variety of renewable technologies.
- Offshore wind energy alone provides nearly half (530 TWh per year) of the total energy in our scenario.
- Biomass (274 TWh per year) and ambient heat (around 105 TWh per year, extracted from ground, water and air by heat pumps) also play major roles. Other contributions are made from solar thermal and geothermal heat (about 40 TWh per year).

What's the problem?

In 2010, around 82% of all of the UK's GHG emissions came from producing energy (DECC, 2013). Burning coal, gas and oil emits carbon dioxide (CO_2). Together these fuels provided around 90% of the UK's primary energy supply, the 'raw' amount of energy supplied before conversion losses. This is illustrated in figure 3.16.

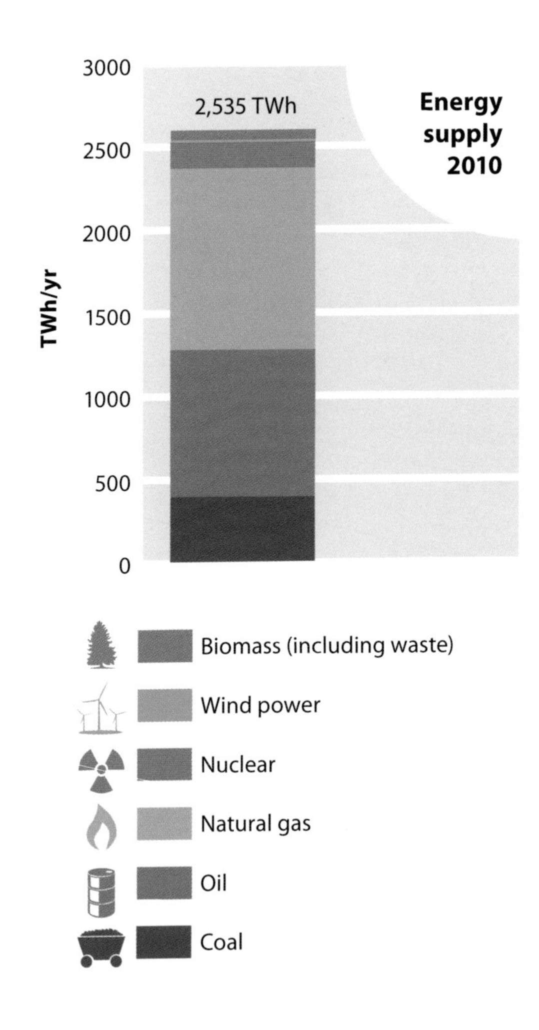

Figure 3.16: Energy supply today (DECC, 2012).

In 2010, renewable energy sources provided 7% of our electricity, and their share of our total energy demand (heat, transport and electricity) was 3.4%. A year later, renewables produced 9% of the electricity and 4.3% of the total energy. This shows that, in relative terms, renewable energy is growing rapidly in the UK, with year-on-year growth of around 20% in recent years. But in absolute terms, renewable energy still currently plays a very minor role compared to fossil fuels. Also, a large proportion of renewable energy generation today comes from burning wood or biodegradable (plant- or animal-based) waste. There are limits to how much we can (or would want to) increase energy production from these sources. If we want to significantly increase the contribution of renewable energy, we need to dramatically increase the role of wind, marine (wave and tidal) and solar energy. In 2011, these sources only supplied around 1% of our total energy, but their contribution is rising sharply. The amount of energy we could theoretically produce from them is enormous.

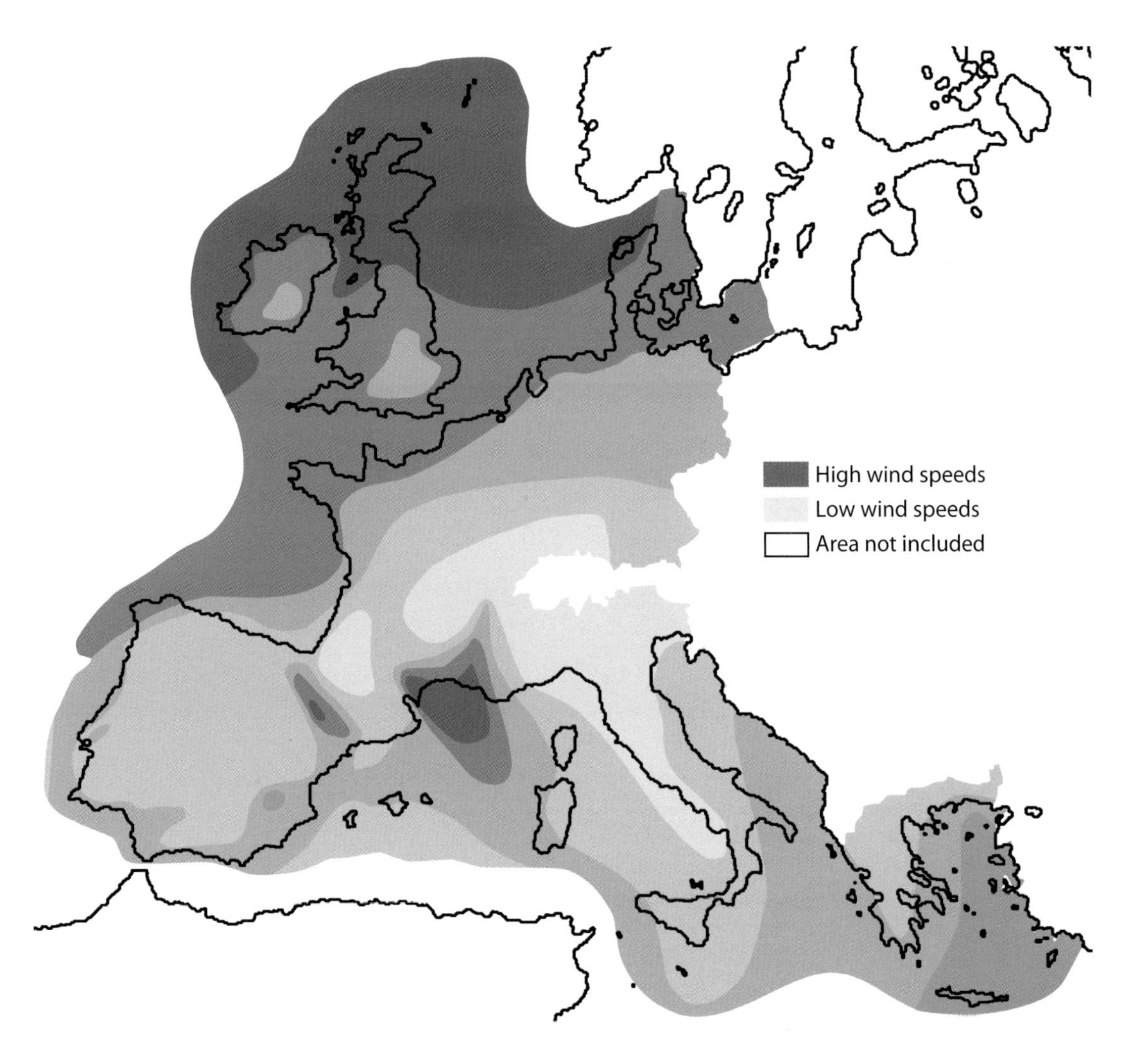

Figure 3.17: European wind speeds at 50 meters above ground level, ranging from the highest (dark blue), to the lowest (light blue). This represents sheltered and open areas, on hills and ridges, coastal areas, and in the open sea, though the highest wind speed and lowest wind speed will be different in each topographical area. Adapted from Troen and Petersen (1989).

What's the solution?

Energy supply from fossil fuels can be replaced with a variety of renewable energy sources that do not emit GHGs. These are:

Wind power

The position of the British Isles as Europe's 'wild and windy' western fringe (see figure 3.17) gives us one of the best wind power resources in the world. What's more, wind power also has the advantage that, statistically, wind speeds are stronger during the winter season when energy demand is highest. This does not mean that the wind always blows when we need energy, but it does show why wind power can help to meet a significant proportion of our energy demand. Currently, most UK wind turbines are installed on land (onshore), but the greatest potential is out at sea (offshore).

© Joanna Wright

Onshore wind: Turbines are easier to install, but as wind speeds are lower over land they produce less energy. The best locations for onshore wind are typically near the coast or on hills. It is estimated that by putting onshore wind turbines in all suitable places, we could produce more than 60 TWh per year (Pöyry, 2011). This is comparable to the amount of energy produced by UK nuclear power stations in 2010, and six times the energy produced from onshore wind turbines in 2011. If we want to make full use of onshore wind power, then we need to accept that wind turbines will become a prominent feature in large parts of the country, including some areas which many people would like to protect from industrial development.

Offshore wind: Out at sea wind speeds are higher. There are also fewer objections to putting very large wind turbines far away from where we live. A 10 MW wind turbine – the kind of size we can expect in a few years – will be as tall as the Gherkin building in London (180 m). A single turbine of this size can produce enough energy for thousands of households, and these machines will likely form the backbone of a future renewable energy system.

Where the sea is relatively shallow – the current limit is depths of 40-60 m – it is possible to build fixed turbines with foundations in the seabed. All existing commercial offshore wind farms are of this type. It has been estimated that the amount of energy we could produce from installing fixed offshore turbines is around 400 TWh per year (Offshore Valuation Group, 2010), more than the UK's current total electricity consumption (320 TWh in 2010). This would require more than 10,000 large fixed offshore turbines. Most of these turbines would be in the North Sea, where very large shallow sandbanks, like the Dogger Bank, could accommodate huge wind farms.

Where the sea is too deep for fixed foundations it is possible to use floating turbines that are anchored to the ocean floor by cables. Full-scale prototypes of this technology have successfully been tested for years. The 65 m tall floating Hywind turbine (with a maximum power output of 2.3 MW) has been operating in 200 m deep waters off the west coast of Norway since 2009, surviving 90 mph winds and

19 metre wave heights. The theoretical potential for rolling out this technology is massive, especially in the deeper waters of the Atlantic off the coast of Scotland and Cornwall. The Offshore Valuation Group (2010) report estimates that we could produce more than 1,500 TWh per year from floating wind turbines alone – this is close to the UK's energy demand in 2010 (1,700 TWh).

Wave and tidal power

Compared to wind power, wave and tidal power generation is still at a very early stage of development. Tidal stream systems resemble 'underwater wind turbines' and produce electricity from natural underwater currents in places such as the Pentland Firth between Scotland and the Orkney islands. Wave power systems produce electricity from waves on the surface of the ocean. According to the Offshore Valuation Group (2010) report, the UK could produce 40 TWh per year from wave power and 116 TWh per year from tidal stream. However, existing wave and tidal stream power projects are still at the prototype stage, and current estimates of their full potential vary greatly. Tidal range projects use barrages or artificial lagoons to produce energy from rising and falling tides. The Offshore Valuation (2010) report estimates that we could produce 36 TWh per year from this technology, with a large contribution (16 TWh per year) from a scheme in the Severn estuary. However, depending on the choice of technology, the local environmental impact of such schemes (for example, reducing tidal range) can be very significant.

Hydropower

Hydropower – generating electricity from water flowing downhill – has a long history in the UK. In fact, the world's first public electricity supply

was from a generator driven by a water wheel in Godalming, Surrey, in 1881. Today, the UK produces around 5 TWh per year from hydropower (DECC, 2012). Most of this is from large power stations, and there is limited scope for building more of these. However, Arup (2011) assume that significant growth in smaller 'micro' hydropower schemes could bring the total production to 8 TWh per year.

Solar photovoltaics (PV) and solar thermal

Solar panels can be used to produce electricity (solar PV) or heat (solar thermal, or 'solar hot water'). South facing roofs are ideal but east or west facing roofs can also be suitable for either technology. The total potential for energy generation is large if all roof areas in the UK are considered; it has been estimated that solar panels on UK roofs could produce 140 TWh of electricity and 116 TWh of hot water every year (DECC (2010) 2050 pathways, level 4). Solar farms in fields could theoretically produce even more energy, but they could compete with other land uses, such as food production. Both solar PV and solar thermal produce much more energy in summer than in winter.

Geothermal electricity and heat

In some parts of the UK, including Cornwall and East Yorkshire, hot rock layers can be accessed by drilling to a depth of several kilometres. The heat can be used to produce electricity in Combined Heat and Power (CHP) stations and to deliver heat to district heating systems that supply hundreds or thousands of households through well insulated heat pipelines. Just how much energy we could produce from geothermal heat in the UK is still debated, with figures of up to 35 TWh per year of electricity (ibid.).

Ambient heat for heat pumps

A heat pump can be seen as a kind of 'heat concentrator' because it takes relatively 'dilute' (low temperature) 'ambient' heat energy from the air, the ground or from (sea or fresh) water, and delivers it as more 'concentrated' (higher temperature) heat. For example, an air source heat pump (ASHP) extracts heat energy from a large amount of cold outside air and uses it to produce a much smaller amount of hot water, which can then be used to heat our homes. Heat pumps need electricity to run but for every unit of electricity input they can deliver two to four units of heat. Today, the overall benefits of heat pumps are often limited, as the electricity they consume is mostly produced in inefficient fossil fuel power stations. But in a future powered by a large amount of wind power, heat pumps are a great way to turn renewable electricity into heat.

Biomass

Plants store energy from the sun in their branches, trunks, leaves and roots. This 'biomass' can then be burned in boilers and power stations to produce heat and electricity. It can also be used to produce biogas and biofuels, or combined with hydrogen to create synthetic liquid and gaseous fuels, as discussed in *3.4.2 Balancing supply and demand* and *3.4.3 Transport and industrial fuels* below. Burning biomass is 'carbon neutral' – no GHGs are emitted overall since the same amount of CO_2 has been absorbed during the plant's growth as is subsequently released during burning. As such, there is no net increase in CO_2 in the atmosphere if a new plant is grown for every plant burned.

The burning of solid biomass in the form of wood has been used to produce energy in the form of heat for millennia. However, there are many competing uses for land in the UK (as discussed in *3.6 Land use*) and this puts a limit on how much we can use for 'growing energy'.

Our scenario

In our scenario, we use a variety of different renewable energy technologies. The energy mix is shown in table 3.1, and relies most heavily on offshore wind power.

The energy flow diagram (figure 3.18) illustrates the production and use of energy in our scenario. It illustrates the central role of electricity – more than 60% of all energy is supplied in this form, compared to less than 20% today. This is in part due to the central role of wind turbines that produce electricity on the supply side, and also to the electrification of

Renewable electricity	Energy (TWh per year)	Details
Offshore wind	530	140 GW maximum power, 14,000 turbines rated 10 MW
Onshore wind	51	20 GW maximum power, 10,000 turbines rated 2 MW
Wave power	25	10 GW maximum power
Tidal (range and stream)	42	20 GW maximum power
Solar PV	58	75 GW maximum power, covering 10-15% of UK roof area
Geothermal electricity	24	3 GW maximum power
Hydropower	8	3 GW maximum power
Total electricity	**738**	
Renewable heat	**Energy (TWh per year)**	**Details**
Solar thermal	25	Covering around 3% of UK roof area
Geothermal heat	15	
Ambient heat	105	Extracted from air, ground and water by heat pumps
Total heat	**145**	
Biomass	**Energy (TWh per year)**	**Details**
For biogas and carbon neutral synthetic gas	94	From waste (37 TWh) and grasses for anaerobic digestion (AD) (57 TWh)
For carbon neutral synthetic fuel	143	From Miscanthus and Short Rotation Coppice (SRC)
For heat	37	From Short Rotation Coppice (SRC) and Short Rotation Forestry (SRF)
Total biomass	**274**	
TOTAL RENEWABLES	**1,157**	

Table 3.1: Energy mix in our scenario.

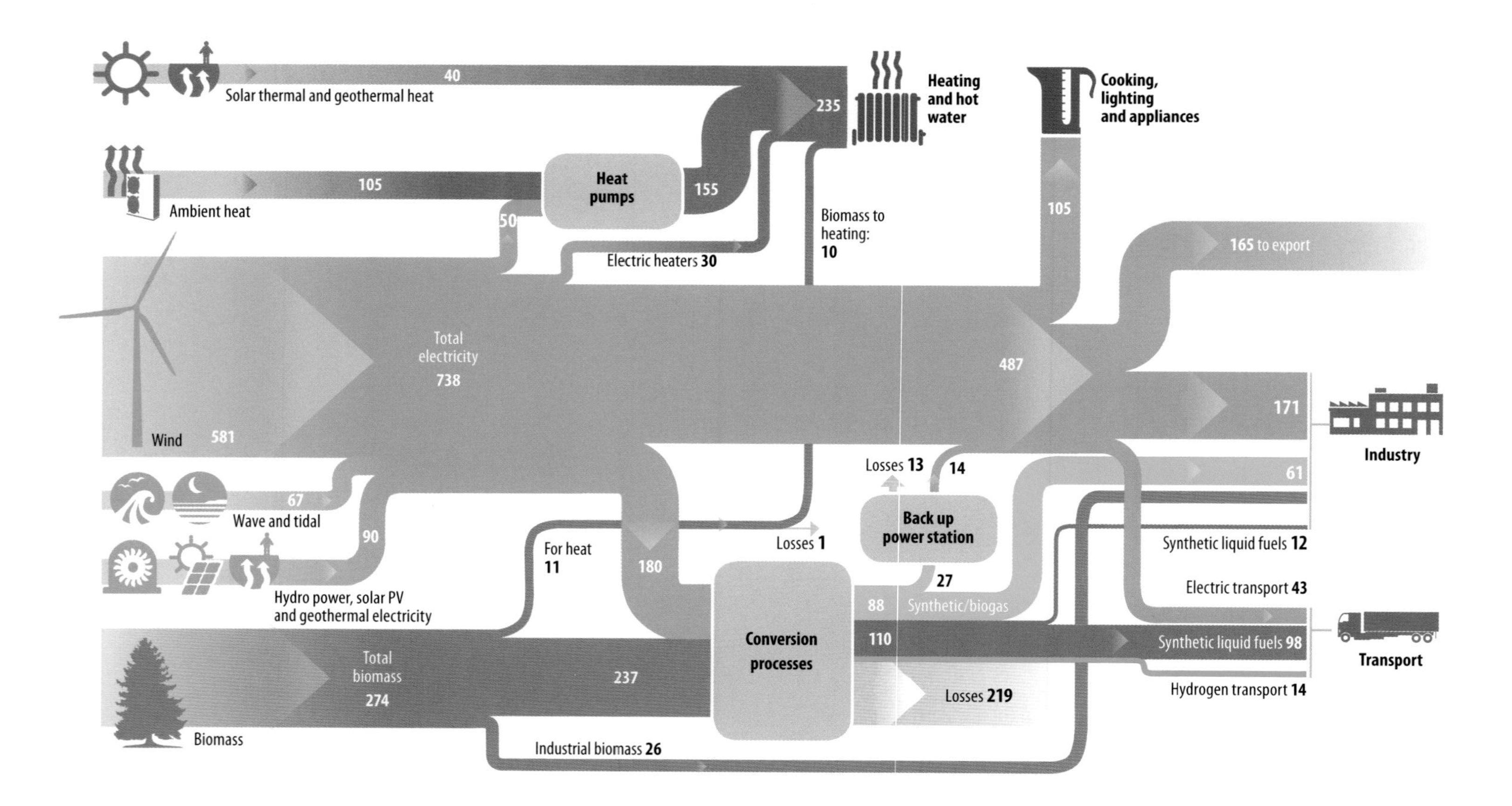

Figure 3.18: Energy flows in our scenario – from supply to demand. Numbers used here are rounded up or down to the nearest TWh and so inputs and outputs may not add up exactly.

heating and transport on the demand side. Biomass also plays a big role in our scenario. As discussed in 3.6.2 *Growing energy and fuel*, this has important implications on land use in our scenario.

3.4.2 Balancing supply and demand

The section describes how we can balance fluctuating energy demand and supply by managing our demand, and creating a back up system with carbon neutral synthetic gas.

Summary

- As most of the energy in our scenario is from variable (fluctuating) sources, there is often a mismatch between supply and demand, with both large surpluses and shortfalls.
- Adding more electricity generating capacity (for example, more wind turbines) would increase surplus electricity production without significantly reducing the problem of shortfalls.
- Shifting certain energy demands to times of high energy supply and combining different renewable sources of energy helps, but it doesn't completely solve the problem.
- Our scenario combines various short-term energy storage mechanisms (hours to days) with the capacity to store up to 60 TWh of carbon neutral synthetic gas for months or years.
- On average, we would be producing 27 TWh of synthetic gas every year, which would be used only as and when required.
- Although overall synthetic gas covers only a very small percentage of our total energy supply, it plays a critical role at times when demand is high and supply from variable renewable sources is low – for example in the cold, windless December of 2010.

What's the problem?

The previous section explains how in our scenario the total amount of renewable energy produced in an average year (about 1,160 TWh) is more than enough to meet the demand (about 770 TWh per year on average). However, as both demand and supply of energy in our scenario are variable (fluctuating) it is still a challenge to make sure that the supply always meets the demand.

Energy demand is variable

The amount of energy we use changes all the time. Currently, our electricity consumption increases rapidly between 5 a.m. and 9 a.m. on a weekday; it reaches its peak in the evening when we come home from work and switch on lights, cookers and televisions. Electricity demand can rise sharply when thousands of kettles are switched on during a TV advertising break or when clouds move over the skies of a big city and lots of people switch on the lights. Also, our demand for heating increases sharply when it gets colder. The distribution infrastructure for gas and liquid fuels has a number of built-in buffers – petrol stations and refineries have large fuel tanks and the gas grid has various stores, including the pipelines themselves. In contrast, the electricity system has much less built-in buffer capacity, hence the supply of electricity always needs to closely match demand. If in the future electricity plays a larger role in heating (heat pumps) and transport (electric cars) then dealing with demand variability will become more challenging.

Renewable energy supply is variable

The energy supply (or 'output') from most forms of renewables is variable. Whereas a nuclear power station might produce the same amount of energy whatever the weather, renewables produce different amounts of energy depending on how fast the wind is blowing, or how much sunshine there is – factors that are beyond our control. With wind power, the changes in energy output can be very sudden. Even with thousands of wind turbines spread around the whole of the UK, it is possible that energy production can near its maximum on one day and be close to zero the next. Moreover, we cannot change these things according to our needs.

This does not mean that renewable energy supply is unpredictable. We can predict the tides centuries ahead, and even predict wind speeds reasonably well a few days in advance. Combining a diverse

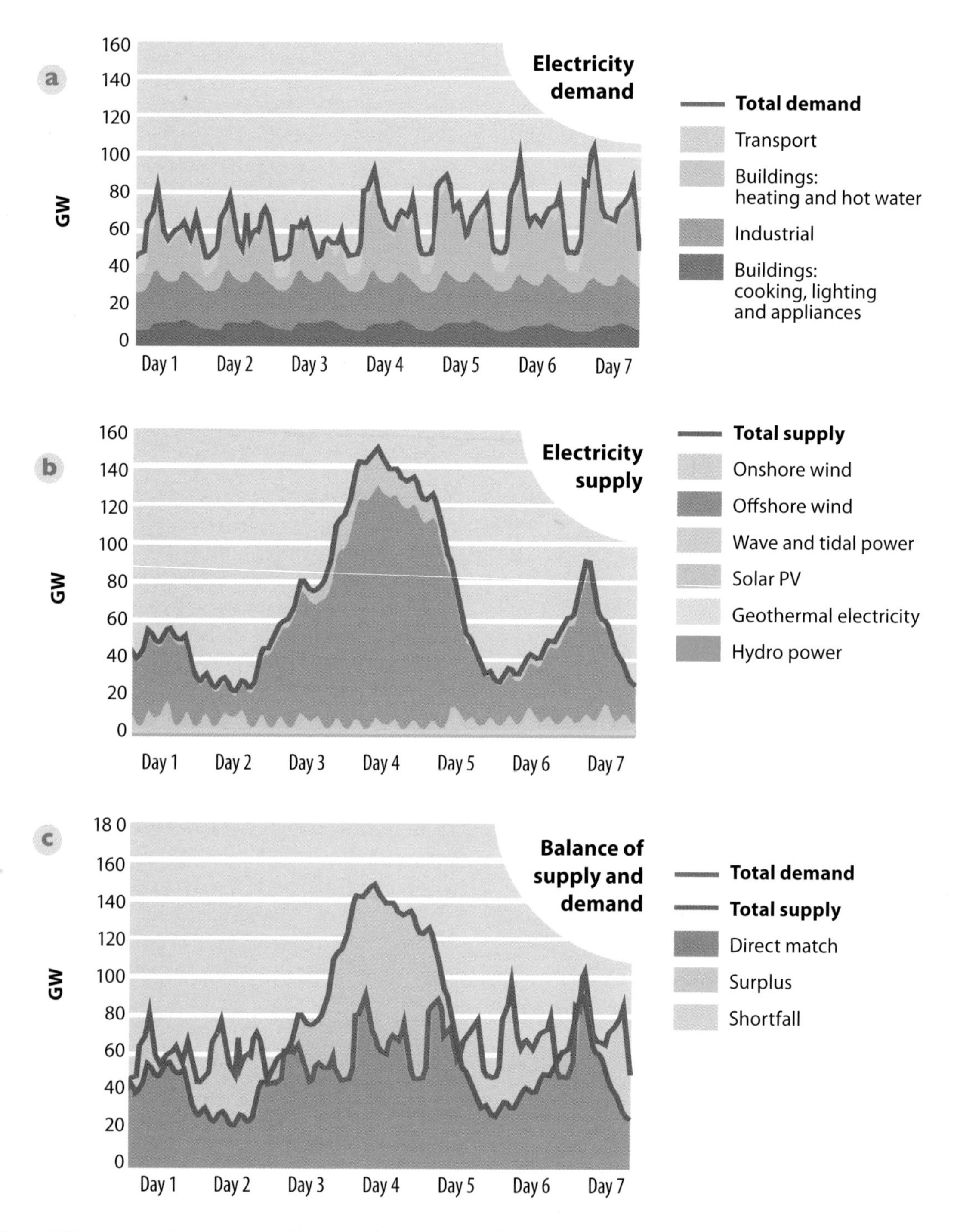

Figure 3.19 a, b, and c: An example of 168 hours (7 days from the 13th – 19th December 2010) of (a) electricity demand, (b) electricity supply, and (c) the balance between them. Supply and demand are modelled using ten years worth of hourly UK weather and electricity demand data. The marked increase in electricity demand for heating during the last four days in (a) reflects colder UK temperatures. Figure (c) shows that at times there may not be enough supply to meet demand (red areas = shortfall), or at other times there may be a greater supply than is needed (blue areas = surplus).

mix of different renewable energy sources can help 'smooth out' energy supply. However, our research shows that even when we combine all the renewable energy sources available in the UK, the energy supply will fluctuate significantly, for example, between a windy, sunny day (lots of energy) and a calm, dark night (little energy). And just adding more generating capacity, for example building more wind turbines or solar panels, is not enough to solve the issue, either. Our calculations suggest that, beyond a certain point, adding more generating capacity will primarily increase the amount of energy that is surplus to requirements without making much difference at times of low renewable energy supply.

Supply does not match demand

Unfortunately, our variable energy demand and variable energy supply don't necessarily 'match-up' – they don't go up and down in step. Figures 3.19a and 3.19b illustrate a typical pattern of electricity supply and demand in winter. A few days of strong winds and waves (lots of energy) are followed by days of calm (little energy). Energy demand also fluctuates – it is typically higher during the daytime, and higher still on cold days because of the demand for heating.

Sometimes renewables supply much more electricity than there is demand for, but at other times wind, waves, tides and solar combined do not produce enough to supply the energy required (see figure 3.19c). Our research shows that there are significant differences over hours, days and even years. For example, 2010 was a year with very cold winters at each end (high heat demand) and unusually low wind speeds (low renewable electricity supply), whereas 2011 was a warmer year with stronger winds. Finding ways to deal with these fluctuations is one of the biggest challenges in powering the UK on 100% renewable energy. We need to ensure our lights stay on and our houses stay warm even during a dark windless night, or during a year with low wind speeds and cold winter months.

What's the solution?

The infrastructure of a renewable energy supply must incorporate some way of 'balancing out' this potential mismatch in supply and demand that is flexible and responsive to fast-changing weather. There are two main methods that can work in conjunction.

Shifting demand to match supply (demand management)

One way to balance supply and demand is to change our energy consumption patterns so that we consume more energy when supply is plentiful, and need less when it is scarce. Industry and some households already pay less for energy during the night when demand is low. It is not difficult to imagine a future in which electricity will be cheaper when it is windy and demand is low, and more expensive when it is calm and demand is high. This could provide an incentive to consume more energy at times when supply exceeds demand and to reduce consumption when energy is in short supply.

'Smart' appliances (such as washing machines and freezers, as well as industrial processes) will automatically run more when electricity is cheap – at times of high supply and low demand – in order to minimise energy consumption when electricity is expensive and in short supply.

'Smart' car charging of millions of electric vehicles could play an important role. Their very large electricity demand can very easily be 'shifted' to times when there is a surplus in the supply of electricity, for example at night or during windy periods.

Storing energy

There are a number of options for storing energy during times of surplus supply so as to make it available at times when more energy is needed. Different types of storage can perform different roles. Sometimes we only need to store energy for short periods – hours or days. At other times, over a very cold and calm winter period for example, we need to be able to build up energy stores for longer periods in advance, in order to make sure we have enough energy to last.

What is crucial for any energy storage solution working with a variable renewable energy supply, is that the 'building up' or the 'emptying' of a store is flexible and, if necessary, relatively quick. We need a dispatchable energy store that can be called upon whenever demand requires it.

For hours or days: There are a number of energy storage options that can help balance out supply and demand over timeframes of a few hours or days.

- **Pumped storage** is used today to store electricity by pumping water uphill into a reservoir at times of surplus energy supply and then letting the water flow downhill through a hydropower turbine when energy is needed. This form of energy storage can be activated very rapidly, but the total amount of energy that can be stored is small. The UK consumes far more than 1,000 GWh of energy on a single cold winter day. The UK's largest pumped storage station, Dinorwig in North Wales, can only store around 10 GWh of electricity.
- **Batteries** in electric vehicles can help shift some electricity demand (as described above). But with today's battery technology, dedicated battery storage – batteries installed exclusively for the purpose of storing surplus grid electricity – is not as cost-effective as some other ways of storing energy.
- **Heat storage** offers an attractive solution in the UK where a large proportion of electricity would be used for heating. Heat can be stored over a few hours or days without significant losses in well insulated hot water tanks (those required, for example, in solar thermal systems). Two hundred litres of storage per household – either individual hot water cylinders, or large external heat stores connected to district heating systems – can store around 100 GWh of heat. This allows heat pumps to play an important role in demand side management as they can be run at times when electricity supply exceeds demand.
- **Hydrogen** can be made by the electrolysis of water – splitting H_2O into hydrogen (H) and oxygen (O) using electricity. Electrolysers can use electricity at times when there is abundant surplus of electricity, to create hydrogen gas for storage. In principle, hydrogen can be stored and then used directly to produce electricity using gas turbines or fuel cells. However, hydrogen is a very light gas that needs to be highly compressed for storage. It is also quite explosive and can even corrode metal. It is possible to store relatively large amounts of hydrogen (a few 100 GWh) over long periods of time, for example in salt caverns. However, compared to natural gas (primarily methane), hydrogen is difficult to store and transport and there is almost no existing infrastructure suitable for it.

What's the difference between *baseload* and *dispatchable* generation?

It is sometimes said that to balance an energy system with a large amount of variable renewable energy you need *baseload* power stations – power stations that produce energy at a constant rate, day and night, such as nuclear power stations. However, constant power output is actually not very useful as it leads to overproduction at times when output from variable renewables is already enough to meet all demand. Instead, our research indicates that there is a requirement for *dispatchable power* – power from generators which can very flexibly increase or decrease output, or even switch off completely, as and when we need them and depending on whether or not there is enough power from variable renewables. Gas power stations, running on either fossil or renewable gas, can be used for this purpose, though of course burning fossil fuel gas emits GHGs.

For weeks or months: Storing enough renewable energy for, say, a cold, dark winter week with low wind speeds is technically very challenging. Realistically, solid, liquid or gaseous fuels are the best option to store the very large amounts of energy required (a few 10,000 GWh). Their high energy densities mean that vast amounts of energy can be stored in relatively small spaces over long periods of time.

Biogas and synthetic gas are both produced from renewable sources. Biogas, a mixture of methane and carbon dioxide, can be produced by anaerobic digestion (AD) – the decomposition of biomass (for example, grass, animal manure or food waste) in an

cenews / Creative Commons ShareAlike 2.0 Generic

oxygen-free environment. Carbon neutral synthetic gas is made via the Sabatier process. Here, hydrogen (made by electrolysis) and carbon dioxide (from burning biomass, or from biogas) are combined to produce methane. Methane is easier to store than hydrogen. The Sabatier process can be seen as 'upgrading' hydrogen to a gas that is easier to handle. The process of using electricity to produce gaseous fuel is sometimes referred to as 'power to gas' (GridGas, 2012).

Methane gas is also the primary component of today's fossil fuel natural gas. The methane in biogas and synthetic gas can be stored in very large quantities just as natural gas is currently. The UK today has a highly developed gas infrastructure that includes storage facilities, such as the Rough gas store off the coast of Yorkshire, which has a capacity of 35,000 GWh. However, methane is a powerful greenhouse gas, so it is very important that any escaping from pipelines or storage is kept to a minimum.

Biogas and synthetic gas, once stored, can be burned in power stations (again, like natural gas today) to provide energy when electricity supply from renewable sources is insufficient to meet demand. Gas power stations burning biogas or synthetic gas can be flexible – we can turn them on or off quickly. We can use them as 'back up' generation to meet demand when electricity supplies from variable renewables fall short. They can also supply

Importing and exporting energy

When planning our scenario we decided to meet all of our energy needs from zero carbon, renewable sources located within the UK, including UK offshore waters. It is important to stress that this is not because we think importing renewable energy from other countries is necessarily a bad idea. It is perfectly possible that solar power from southern Europe or even northern Africa could complement UK wind energy. This is often discussed in the context of a European high voltage 'super grid' which would enable the distribution of large amounts of electricity over long distances with low losses.

However, when designing an energy scenario that allows imports, it is difficult to decide what would be our 'fair share' of foreign renewable energy sources. Crucially, this is true even in a scenario where the UK is a *net exporter* of energy, that is, a country that sometimes has to buy energy but overall sells more energy than it buys. The problem is that other European countries are likely to be in a very similar position to the UK, with low electricity supply when wind speeds are low over the North Sea, and high electricity demand on cold, dark winter days. Therefore, if the UK were to rely on imports for days when its own renewable sources did not produce enough, it would likely find itself competing with these other countries over resources, such as solar electricity from the Mediterranean region.

Without detailed modelling of energy flows for all of Europe we cannot simply assume that our neighbours will want, or be able, to sell us energy whenever we need it. Conversely, it is possible that at times when we produce more energy than we need, our neighbours will also have more than enough energy and would not be willing to pay a high price for our surplus. Therefore, while in our scenario a fairly large amount of surplus electricity (165 TWh per year) is exported, this does not necessarily mean large income from electricity sales.

All this is not to say that energy imports and exports should not play a role in zero carbon energy scenarios. The benefits from exchanging renewable energy with our neighbours could significantly reduce the cost of storage and back up. We are looking forward to working together with researchers from other countries to model energy flows in a 'zero carbon Europe'.

industry for very energy intensive processes which would be difficult to run on electricity (see *3.3.1 Buildings and industry*).

It is important to remember that burning methane is only carbon neutral when it is produced using biomass and/or renewable electricity. When methane gas is produced from biomass, the amount of CO_2 released by burning it is reabsorbed when new biomass plants are grown, resulting in no net increase of GHGs in the atmosphere. Synthetic gas is carbon neutral when the hydrogen used is produced using renewable electricity, and the CO_2 used is from non-fossil fuel sources (like biomass).

The processes involved in creating a significant biogas and synthetic gas back up system have many losses associated with them. As energy is converted between forms (electricity and biomass to gas, and back to electricity), we lose energy in the process – about 50%. However, the ability to store energy in this way forms an integral part of an energy system powered by renewables, and is a good way of using electricity which would otherwise be surplus to requirements.

Our scenario

In developing our scenario, we used real hourly weather data (solar radiation, wind speeds, temperatures, etc.) for the last ten years – a total of 87,648 hours – to simulate patterns of supply and demand. In other words, we looked at how well the technical solutions we propose for a zero carbon future would have fared hour-by-hour under the weather conditions observed in the past decade.

In our scenario:

- **82% of the time**, the supply of renewable electricity exceeds the direct demand for electricity (including electricity for heating and transport) required at any one moment. Due to the very large number of wind turbines and other renewable electricity producers, almost half of the total electricity produced (about 354 TWh per year) is surplus to what is directly required at the time of production. However, 18% of the time, electricity supply

© Joanna Wright

does not fully meet demand.

- **Short-term storage** mechanisms, such as pumped electricity storage (25 GWh storage capacity), 'shiftable' demand from smart appliances and electric car charging (25 GWh), and heat storage (100 GWh heat) reduce the proportion of time during which electricity supply does not meet demand from 18% to 15%. This reduces the amount of surplus electricity to about 345 TWh per year. Crucially, by 'capping the peaks' of unmet demand, these mechanisms significantly reduce the back up power station capacity required (see below). So short-term storage reduces not only the number of hours during which back up is needed, but also the number of gas power stations required.
- **Electrolysis** units, with a maximum power consumption of 35 GW, use around half (180 TWh per year) of the surplus electricity (the rest is exported). The hydrogen produced (126 TWh) is stored mostly in large underground caverns with a capacity to store 20,000 GWh of gas. A small proportion of this hydrogen is used as fuel for hydrogen vehicles (11%) but most of it is used to produce carbon neutral synthetic gas (35%) or synthetic liquid fuels (54%), as explained below.
- **Biogas and carbon neutral synthetic gas** are burned in gas power stations to supply electricity during the 15% of the time when electricity demand would otherwise exceed supply. In our scenario, we need to produce on average 27 TWh of biogas or synthetic gas as back up every year, to be used as and when required, which in turn produces an average of 14 TWh of electricity per year. We incorporate a large number of (renewable) gas power stations (45 GW maximum output, comparable to the capacity of all gas power stations we have today), but these power stations are inactive most of the time, turned on only when electricity demand

would otherwise exceed supply. Overall, these gas power stations only produce 3% of the electricity in our scenario. But our simulation shows that in weather conditions such as those experienced in December 2010, with very low temperatures and very little wind, such back up power stations would play a critical role, supplying more than half of all electricity on some days. To store enough biogas and synthetic gas for these periods, our scenario includes 60,000 GWh of methane gas storage. Today the UK already has one gas storage facility with a capacity of 35,000 GWh.

3.4.3 Transport and industrial fuels

In the section, we describe how we can provide carbon neutral synthetic liquid fuel to meet transport and industrial energy demands.

Summary

- In our scenario most energy (404 TWh per year) is used in the form of electricity but planes and some commercial vehicles work better with liquid fuels. Even with reduced amounts of travel, they require a total of 98 TWh of liquid fuel and 14 TWh of hydrogen per year.
- There is also a demand for liquid fuel (12 TWh per year) from industry. And gas is required for industry and for long-term energy storage and back up – 61 TWh and 27 TWh per year respectively.
- We use processes that produce carbon neutral synthetic liquid fuels and synthetic gas by combining biomass and hydrogen.
- For these processes, a total of 126 TWh of hydrogen is produced using surplus electricity every year. 14 TWh of this is used directly in transport. 68 TWh of hydrogen is combined with 143 TWh of energy in the form of 'woody' biomass to make the required 110 TWh of carbon neutral synthetic liquid fuels for transport and industry. The remaining 44 TWh of hydrogen, with an additional 94 TWh of biomass, provides the required 88 TWh of carbon neutral synthetic gas and biogas required as back up and for industry.
- There are significant losses in these conversion processes (about 50%), which mean more energy must be put in than we get out. However, it is the form of the fuel and our ability to use surplus electricity that is important here.

What's the problem?

As described in *3.3.2 Transport*, although much of our transport can be electrified, there are some transport needs that can't be met by electricity. Liquid fuels, such as the kerosene, diesel and petrol we use today, offer a much higher energy density – smaller and lighter ways to store energy – than even the best batteries available today. If we want planes and heavy commercial vehicles (such as Heavy Goods Vehicles (HGVs), tractors and diggers) in a zero carbon future, we need to find ways to provide transport fuels with similar energy densities that are carbon neutral and can be produced from renewable energy.

There are also industrial processes that currently use natural gas or liquid fossil fuels, and these processes, too, will require carbon neutral, renewable alternatives.

What's the solution?

There are processes which allow us to produce liquid or gaseous fuels from renewable sources, replicating the fuels we use today but without the associated GHG emissions.

Hydrogen

Hydrogen (produced through electrolysis as described in *3.4.2 Balancing supply and demand*) can also be used to power hydrogen cars. However, the problems that apply to hydrogen storage also apply to using it to power vehicles: hydrogen is difficult to store and transport and, in practice, it would be difficult to use it as the main source of transport fuel. Doing so would require us to develop a whole new infrastructure.

Biofuels

Biomass can be used to produce liquid fuels very similar to today's fossil fuels. **First generation biofuels** are liquid fuels such as 'corn ethanol' or 'rapeseed oil biodiesel' that are produced from biomass in wheat, corn, sugar crops and vegetable oil. They have come under much criticism because their production can require a lot of energy, pesticides and fertiliser. They also grow best on cropland that is often in short supply, and so can compete with food production, or can contribute to land use change and deforestation, mainly overseas. **Second generation biofuels** allow the production of fuels from biomass in more 'woody' plants, such as fast-growing trees and grasses (*3.6.2 Growing energy and fuels*). These can be grown using less fertiliser and on lower quality land not usually used for food crops. However, there are still many competing uses for land in the UK (as discussed in *3.6 Land use*) and this puts a limit on how much we can use for fuel production.

Biofuels from the sea?

Algae (seaweed and microalgae) are currently being researched as a promising alternative to terrestrial biomass (trees and grasses) for carbon neutral synthetic liquid fuel or biofuel.

High biomass yields are possible, and no land is needed to produce them. Seaweed cultivation can be integrated with fish farms, improving the water quality; or even in offshore wind farms (Hughes et al., 2012; Roberts and Upham, 2012). Microalgae can be used in wastewater treatment, reducing the need for chemicals. It also recovers otherwise 'wasted' nutrients contained in the water (Kumar et al., 2011).

But the drawbacks today are also important: harvesting them is more difficult than harvesting terrestrial plants, and the species that are most suitable for producing biofuel and synthetic liquid fuel grow best at between 15°C and 26°C (ibid.), meaning that often they are more suited to the warmer climates of southern Europe than the UK.

Finally, this way of producing biomass for liquid fuels is still at an early stage of research and development. For these reasons it is not included in our scenario.

Carbon neutral synthetic liquid fuel

Similar to the production of synthetic gas, it is possible to produce synthetic liquid fuels by combining carbon, produced from biomass, with hydrogen, produced through electrolysis. The Fischer-Tropsch (FT) process is a collection of chemical reactions that can be used to combine carbon monoxide (which can easily be produced from 'woody' biomass) with hydrogen to form carbon neutral synthetic fuels for heavy commercial vehicles and planes. This combines the advantages of hydrogen (use of surplus electricity) and second generation biofuels (high density liquid fuels from 'woody' biomass).

Just as with synthetic gas, the resulting fuels are carbon neutral: the CO_2 emitted by burning them was initially taken in by the biomass as it grew, and the electricity used is renewably produced. Over the long-term there is no net increase in greenhouse gases in the atmosphere.

From an energy perspective, the conversion of surplus electricity (via hydrogen) and biomass into liquid fuels is not very efficient, as more than half of the energy is lost in the process. However, it is the *form* of energy that is important here – liquid fuels allow us to do things (fly planes, drive heavy commercial vehicles) that would otherwise not be possible.

Our scenario

In our scenario, every year, 180 TWh of surplus electricity is used to produce 126 TWh of hydrogen through electrolysis. 14 TWh of this hydrogen is supplied directly for hydrogen vehicles every year. 68 TWh of the hydrogen and 143 TWh of energy in the form of 'woody' biomass (see *3.6.2 Growing energy and fuel*) are combined in the FT process to deliver 110 TWh of carbon neutral synthetic liquid fuels, which are used in aviation (39 TWh), heavy

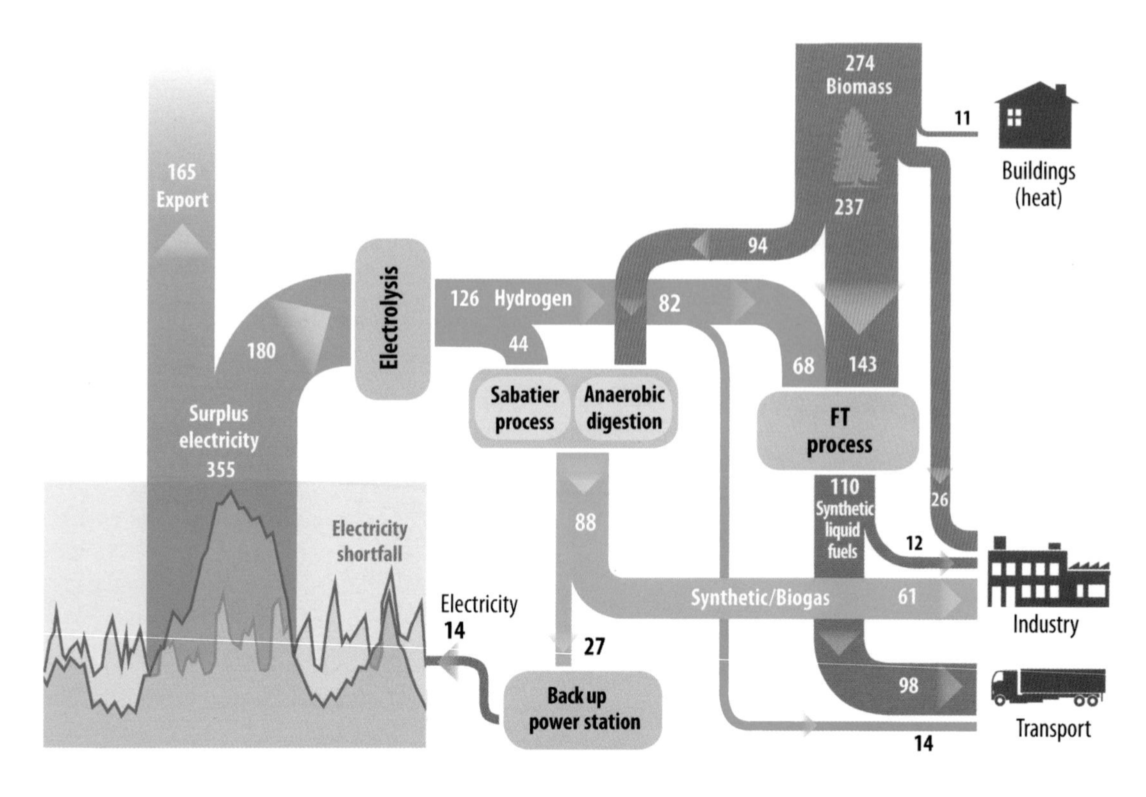

Figure 3.20: From surplus electricity and biomass to synthetic fuels for industry, transport and energy system back up. Losses are not shown in this figure.

commercial vehicles (59 TWh) and industry (12 TWh).

The remaining 44 TWh of hydrogen, together with 94 TWh of biomass (mostly grasses – 57 TWh – and some waste – 37 TWh (*3.5.2 Waste*)) provide 88 TWh of biogas and synthetic methane. 61 TWh of this is used by industry, and 27 TWh (as described in *3.4.2 Balancing supply and demand*) is used as back up to balance supply and demand. Figure 3.20 summarises these processes of producing synthetic fuels for industry, transport and energy system back up.

The large amount of biomass (200 TWh, excluding waste), and therefore land, required is the main limiting factor in the production of synthetic liquid fuels (and hence the amount we can fly, or supply fuel for heavy commercial vehicles – see *3.3.2 Transport*). 'Boosting' fuel production by adding hydrogen from surplus electricity reduces the amount of biomass required. However, even with the use of hydrogen, the amount of land needed to meet today's aviation fuel demand from carbon neutral sources that rely on biomass is likely to exceed the land area of the UK.

Southern Methodist University, Central University Libraries / Public domain

There are some GHG emissions that are not caused by the combustion of fuels for energy. Instead, they occur from the expansion of urban areas, chemical reactions in industrial processes, the leakage of GHGs in industry, businesses and households, and from waste management.

By changing industrial processes and substituting gases and/or products with less polluting alternatives, we can reduce the emissions from businesses, industry and households fairly significantly, but not entirely.

Furthermore, with some changes to the way we deal with waste, it is possible to turn waste processing from a net GHG emitter to a method of capturing carbon. Additional benefits from doing this include energy generation and use of certain wastes for better fertilisation of soils.

Non-energy emissions summary:

- Emissions from non-energy sources accounted for just over 8% of UK GHG emissions in 2010 – 54.4 $MtCO_2e$. These came from urban expansion, industrial processes, leakage of some GHGs in industry, businesses and households (for example in gas pipelines), and from waste management – mainly landfills.
- These emissions are reduced to about 21 $MtCO_2e$ in our scenario – a 61% reduction. However, using technologies available today, it is not possible to completely eliminate these emissions.

3.5.1 Industry, businesses and households

In this section, we describe ways of reducing non-energy emissions from industry, households and business.

Summary

- Non-energy emissions from industry, businesses and households together accounted for just under 6% of total UK GHG emissions in 2010 – 38 $MtCO_2e$. Pre-recession, when industrial output is likely to have been higher, there were more emissions from these areas – 41 $MtCO_2e$
- In our scenario, these emissions are reduced to just under 16 $MtCO_2e$ – by changing industrial processes and substituting gases and/or products with less polluting alternatives.
- There is potential for the complete elimination of emissions from iron and steel production but the methods are as yet unproven.

What's the problem?

In addition to the GHG emissions from burning fossil fuels for energy, GHGs are emitted by chemical reactions in industrial processes. GHGs can also leak directly into the atmosphere from products containing them, or when they are moved around, and there are some emissions associated with the expansion of urban areas. In total, the non-energy emissions from industry, businesses and households accounted for 6% of total UK GHG emissions in 2010 (DECC, 2013; DEFRA, 2013).

Non-energy emissions specifically from industry, businesses and households can be divided into six categories:

1. Iron and steel production: CO_2 emissions are incurred in iron and steel manufacture when carbon is used to reduce iron oxides.
2. Cement production: CO_2 emissions are incurred in the production of clinker, a component of cement, when limestone ($CaCO_3$) is converted to lime (CaO).
3. Emissions from making fertiliser and synthetic materials: emissions occur from the chemical reactions involved in making these products.
4. Leakage of 'super greenhouse gases': around 2% of UK GHG emissions are super greenhouse gases (super GHGs) (DECC, 2013). They are released from refrigeration and aerosols, and during foam manufacture. Although only released in tiny quantities, they are very powerful (between 150 and 23,900 times as powerful as CO_2 (ONS, 2012)) and so make a significant contribution to UK GHG emissions.
5. Leakage of methane (CH_4): this occurs from the current gas network and from disused coal mines.
6. Urban expansion: this causes GHG emissions from soils and plants as they are cleared for development and was responsible for emitting roughly 6.2 $MtCO_2e$ in 2010.

What's the solution?

There are various ways these emissions can be reduced or eliminated completely:

1. Total UK emissions from iron and steel production could be brought down by around 80% by 2030 (AEA, 2010). This could be achieved by: reusing and recycling more steel; powering more iron and steel production with electric arc furnaces; using biomass, biogas and carbon neutral synthetic gas for heat; and using 'top gas recycling' to recirculate gases so that more carbon is fully oxidised. However, this still leaves some emissions from the reduction of iron oxide using carbon. There may be ways to completely eliminate these emissions (see box on page 75), but they are as yet unproven.
2. The substitution of up to 40% of clinker with non-emitting alternatives in cement production is considered feasible, and would achieve an equivalent reduction in emissions (ibid.).
3. Nitrous oxide emissions from producing adipic and nitric acid (used in nylon and fertiliser manufacture, respectively) can be virtually eliminated by changes to how they are made (Lucas et al., 2007).

4. In most cases, it is possible to substitute super GHGs with gases that have low or no greenhouse effect – this could achieve emission reductions of up to 80-90% by 2050 (ibid.). Reductions of 75% should be feasible by 2030.
5. Using less methane and improving network maintenance can reduce methane leakage from the gas network.
6. Halting or slowing urban expansion could decrease emissions – redeveloping, renovating and retrofitting old unused buildings and developing under-occupied areas in urban landscapes offer alternatives.

Our scenario

Table 3.2 below shows the extent to which non-energy emissions are reduced in our scenario given the measures detailed above. However, total emissions also depend on changes in demand for the products causing the emissions. A 2007 baseline is used, as we expect industrial output per person to be

An end to emissions from iron and steel manufacture?

- Iron and steel manufacture could use electrolysis, not carbon, to reduce iron oxide. This would completely avoid CO_2 emissions. In electrolysis, iron ore is dissolved at high temperatures. When electricity is passed through the solution, oxygen and liquid iron are produced. This process has been shown to work on a small-scale (ULCOS, 2010b).
- Another possible carbon neutral way to reduce iron oxide is to use charcoal derived from biomass. It is under investigation whether this could provide a suitable alternative (ULCOS, 2010a).

Whilst these alternatives are promising, neither is sufficiently well proven to be included in our scenario. In addition, Carbon Capture and Storage (CCS) could be used to prevent the release of emissions from iron and steel production into the atmosphere. However, as with CCS in electricity generation, we do not consider it for our scenario (see *3.1 About our scenario*).

Source	2007	Our scenario	
	$MtCO_2e$	% of 2007	$MtCO_2e$
Iron and steel production	5.2	58%	3.0
Cement production	6.1	61%	3.7
Super GHGs	14.0	25%	3.5
Nitric and adipic acid production	2.8	4%	0.1
Other process emissions (from aluminium, lime, soda ash, fletton brick and the production of other chemicals)	1.8	100%	1.8
Leakage of methane from gas network	4.6	15%	0.7
Emissions from disused coal mines	0.5	80%	0.4
Conversion to urban land	6.2	44%	2.7
Total	**41.2**	**39%**	**15.9**

Table 3.2: Summary of non-energy emissions from industry, businesses and households in 2007, representing pre-recession industrial activity (DECC, 2013; DEFRA, 2013), and in our scenario.

at around 2007 levels (that is, pre-recession) in our scenario. The following additional assumptions are made:

- Using a greater proportion of plant-based building materials, for example wood (see *3.6.3 Capturing carbon*), means demand for steel and cement in building construction decreases. However, demand will also *increase* to build wind farms and other infrastructure. Therefore, it is assumed that UK iron and steel and cement production remains at 2007 levels.
- The demand for some products that currently use, and potentially leak, super GHGs will increase. For example, the number of heat pumps, which use refrigerant gases, will increase. However, these products can be switched to gases with a much lower greenhouse effect, so in total a 75% reduction of super GHGs is still considered achievable in our scenario.
- Methane leakage from the gas network is assumed to remain at the same percentage of total gas used as in 2010. However, the synthetic gas used in our scenario is much less than current natural gas use. Methane leakage from coal mines is assumed to continue at the 2010 level.
- Emissions from the expansion of urban areas is reduced by renovating existing buildings and developing under-occupied urban areas.

3.5.2 Waste

This section covers non-energy emissions from waste management processes and describes ways in which they can be reduced.

Summary

- Emissions from waste management contributed about 2.5% to UK GHG emissions in 2010. These mainly come from landfill, but also from waste incineration and wastewater processing.
- Landfill emissions are, however, decreasing due to concerted efforts to divert waste elsewhere (for example, recycling and composting) and increased efforts to capture methane emissions for the production of energy.
- The best way to reduce emissions from waste is to produce less. Consuming less, reusing, recycling

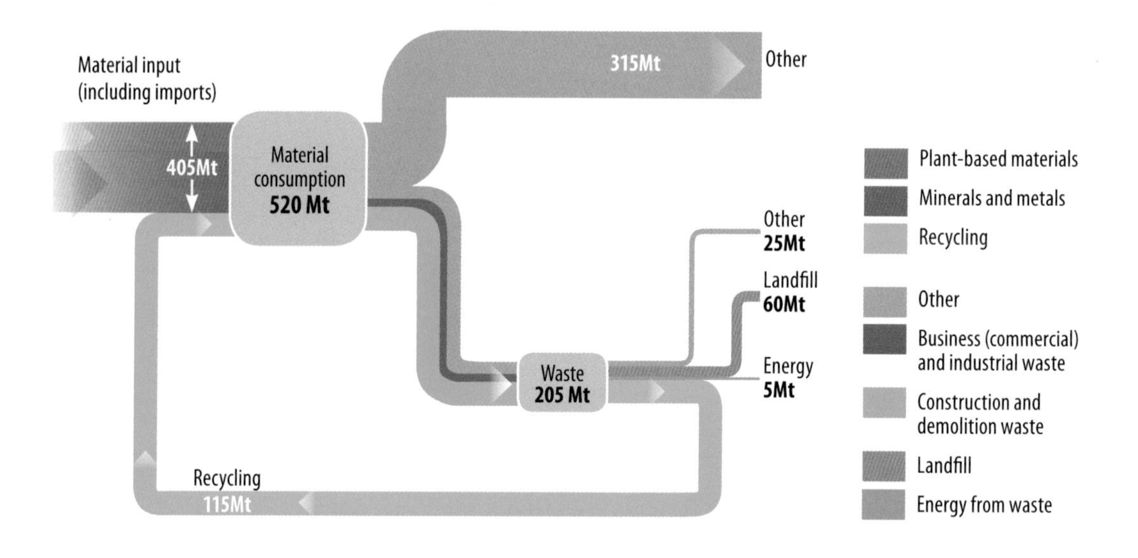

Figure 3.21: Where our waste currently goes in the UK. Adapted from DEFRA (2011b).

and recovering materials and energy are all preferable to putting materials in landfill.

- Landfill can be converted into 'silo storage' units or bioreactors, and wastewater processing plants can be fitted with anaerobic digesters – both of which reduce emissions and produce energy.
- Overall, emissions from waste management in the UK can be reduced by almost two-thirds, to just over 5 $MtCO_2e$.

What's the problem?

In 2010, according to UK statistics, waste management was responsible for 16.5 $MtCO_2e$ (2.5% of the UK's total GHG emissions). The majority of this, 14.7 $MtCO_2e$, came directly from landfill. Wastewater processing (the cleaning of wastewater before it is pumped into rivers and seas) and burning waste contributed 1.5 $MtCO_2e$ and 0.3 $MtCO_2e$ respectively (DECC, 2013).

Only about 22% of the 520 million tonnes (Mt) of products and materials that we consume in the UK every year is recycled (WRAP, 2013), though this is increasing (DEFRA, 2011b). About 60 Mt is landfilled every year (see figure 3.21), and this figure is decreasing (DEFRA, 2011a). We currently waste about 30% of all the food we produce (FAO, 2011).

There are substantial gaps in our knowledge about waste because not all of it is regulated or recorded (Fawcett et al., 2002). Figure 3.22 shows the proportions of waste from some sectors in the UK. In general, the amount of waste produced is thought to be decreasing (DEFRA, 2011b).

Many products have an environmental impact simply from the process of manufacture – in other words, in the extraction of the basic materials, plus GHG emissions from processing and manufacturing. Products can also contain materials which are in relatively short supply globally. By not reusing or recycling these materials we rapidly use up remaining resources.

Any plant-based materials (wood, paper and food, for example) that end up in landfill emit GHGs as they decompose. Since there is very little, or no, oxygen in landfill, these materials don't decompose completely. Some carbon stays in the materials

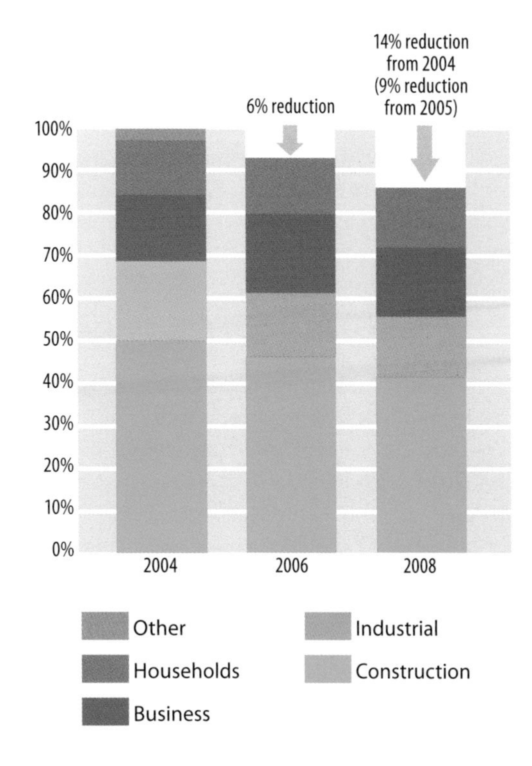

Figure 3.22: Estimated percentage of waste generated by each sector, and demonstration of the reduction in waste generation in recent years. Adapted from DEFRA (2011b).

almost indefinitely, whilst some is released as methane (CH_4), which is a much more powerful GHG than the CO_2 that these plants originally captured. Less 'woody' plant-based materials – food waste, grasses, agricultural and crop residues, decompose relatively quickly and more completely, releasing lots of CH_4, and storing relatively little carbon. More 'woody' materials (timber) decompose less, meaning less CH_4 emissions and more carbon stored per tonne landfilled (UNEP, 2010).

Methane from landfill can be captured and used to produce energy. GHG emissions from landfill have decreased significantly over recent years because of methane capture, and because we are diverting wastes from landfill. GHG emissions from landfill fell 59% between 1990 and 2007 (Environment Agency, *undated*).

What's the solution?

The waste hierarchy

Current government policy and recommendations by the United Nations Environmental Programme (UNEP) for reducing the environmental impact of waste are shown in figure 3.23. Preventing waste should be the first and foremost measure taken (UNEP, 2010).

Benefits from preventing waste, reusing materials and recycling are much greater than those from any waste treatment, even if energy is recovered in that process (ibid.). Wasting less would mean consuming less, also resulting in less manufacturing. In turn this reduces the environmental impact (and GHG emissions) from production and manufacturing processes, and from waste treatment; to change only what we do with waste once it is generated has no impact at all on the emissions in the production and manufacturing stages.

Recycling 70% of household waste could save 4.4 $MtCO_2e$ per year, for example (Fawcett et al., 2002), and there are many, many more opportunities to recycle in business and industry. As many things should be reused or recycled as possible (Michaud et al., 2010), though burning some materials, like medical wastes, may be the only way to prevent potentially dangerous contamination.

One important precondition for reducing emissions from waste is to sort it into different types, so that it can be treated appropriately. This applies to non-plant-based materials (plastics, metals, etc.) though these are not the major contributors to emissions from landfill. Plant-based biodegradable materials that decompose, contributing to landfill emissions, could be sorted as follows:

- Food and agricultural waste (high GHG emissions in landfill, low carbon storage potential) should not be landfilled. There are better purposes for food waste, if we are careful – for example, feeding livestock, or creating compost for soils. Agricultural waste (manure from livestock, agricultural or crop residues, animal industry wastes) can be used to produce energy through anaerobic digestion (AD). The residue from AD still contains all the nutrients in the original material and so can be reapplied to soils as compost or fertiliser (UNEP, 2010).
- More 'woody' waste (off-cuts from forestry, branches, bark and sawdust), could either be used to make biochar via pyrolysis or as biomass for energy production (ibid.) (see *3.4 Power Up* and *3.6.3 Capturing carbon*).

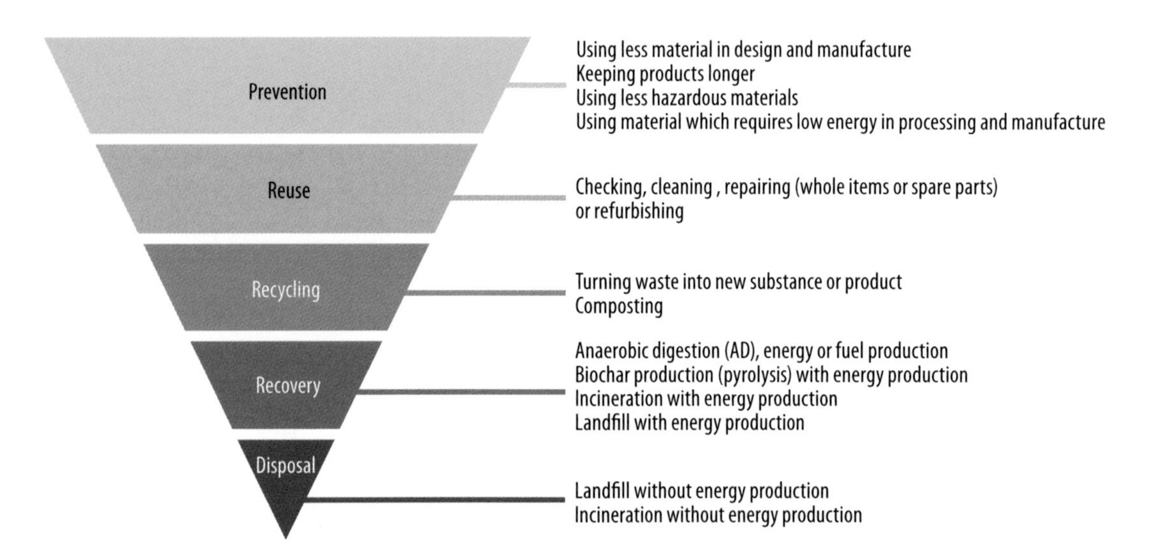

Figure 3.23: The waste hierarchy. Measures at the top of the triangle are best. Adapted from DEFRA (2011b).

Pylon757 / Creative Commons ShareAlike 2.0 Generic

- There are many opportunities for reducing, reusing and recycling plant-based construction and demolition materials (up to 90% of all waste from construction and demolition is recycled in some countries (Symonds, 1999)). Though, eventually, waste will occur (for example, if wood products become partially rotten or are damaged beyond reuse or repair). This waste can also be converted into biochar (3.6.3 *Capturing carbon*), but if it contains materials that have been heavily treated with chemicals, or if it is likely to produce harmful residues when burned, landfill perhaps remains the only option.

Better design and protection of landfill sites – for example, covering waste within a few months to stop decomposition – can create 'storage silos' that capture carbon (UNEP, 2010; Hogg et al., 2011). Alternatively, promoting decomposition by adding air or water can create 'bioreactors' that produce energy. In both cases there is the potential to eliminate nearly 100% of the methane emissions from landfill (ibid.).

Wastewater processing

All sewage and wastewater treatment plants could be fitted with anaerobic digesters (ADs), using the gases to produce energy (biogas), while enclosing tanks and adding waste gas scrubbing mechanisms could further reduce emissions (AEA Technology Environment, 1998b).

Our scenario

Most of the plant-based waste streams in our scenario are diverted from landfill to other uses:

- Food waste is halved and we assume the remaining portion feeds livestock (pigs) or is composted.
- All biodegradable agricultural waste (straw from cereals, for example), waste from sewage systems, poultry waste and manure from livestock is used to produce energy through anaerobic digestion (AD) of the biomass. The residue is used as compost or fertiliser on agricultural land and land used to grow energy crops.

- The amount of 'woody' construction and demolition waste increases in this scenario due to planting new forests and using more plant-based materials in buildings (see 3.6.3 *Capturing carbon*). It is assumed that about two-thirds of all construction and demolition waste (once it has been reused and recycled) will not be safe to turn into biochar, and will be landfilled. New landfills are built as 'storage silos' meaning a negligible amount of methane is released, and methane capture from existing landfills is improved.

What about non-plant-based materials?

Most manufacturing processes, and hence waste streams, are not explicitly modelled in our scenario. Reuse or recycling of any non-plant-based materials (like metal, plastics and glass) are assumed to contribute to energy demand reductions from industry if they are produced in the UK (see *3.3.1 Buildings and industry*).

Therefore:

- There is a 75% reduction in emissions from landfills (91% reduction from 1990 levels are assumed feasible by AEA Technology Environment (1998a)).
- Emissions from burning waste are assumed to remain the same.
- Methane emissions from wastewater processing are used to produce energy, and N_2O emissions are reduced by 25%.

Together, these measures mean the waste sector in 2030 emits 5.1 $MtCO_2e$ – just over a third of 2010 emissions.

Biogas from anaerobic digestion of some biodegradable waste and wastewater processing, and a small amount of methane from remaining landfills, help meet some of our energy demands in our scenario. Together, they produce the equivalent of 37 TWh of biomass for biogas production (see 3.4 *Power Up* and 3.6.2 *Growing energy and fuel*).

The previous sections – 3.3 *Power Down*, 3.4 *Power Up* and 3.5 *Non-energy emissions* – show that most of the UK's emissions (about 90%) can be reduced significantly – almost to zero, save a few industrial and waste management processes that still emit GHGs. The remaining impact on climate change from these areas is about 28.2 $MtCO_2e$ per year in our scenario – 15.9 $MtCO_2e$ from non-energy emissions from industry, businesses and households, 5.1 $MtCO_2e$ from waste management, and 7.2 $MtCO_2e$ from the effects of aviation (see 3.3.2 *Transport*).

However, there are still emissions associated with agricultural food production, and those from land use changes and land management practices – about 10% of current UK emissions. We will see how we can reduce some of these emissions in this section.

That said, since our target is *net* zero for all emissions, this is still not quite enough.

There are therefore now two new demands on land in our scenario, aside from food production. One is the need for biomass – to fuel some parts of our transport system, and to provide back up for our energy system. The other is to 'balance' the impact of our remaining emissions by capturing carbon – removing CO_2 from the atmosphere every year in equal measure to this impact. In doing this, the UK will essentially be cleaning up its own mess within its own territory.

This is the last item in the jigsaw. Our use of land in the UK will provide food, energy resources and carbon capture, which allows the UK to be truly net zero carbon.

Land use summary:

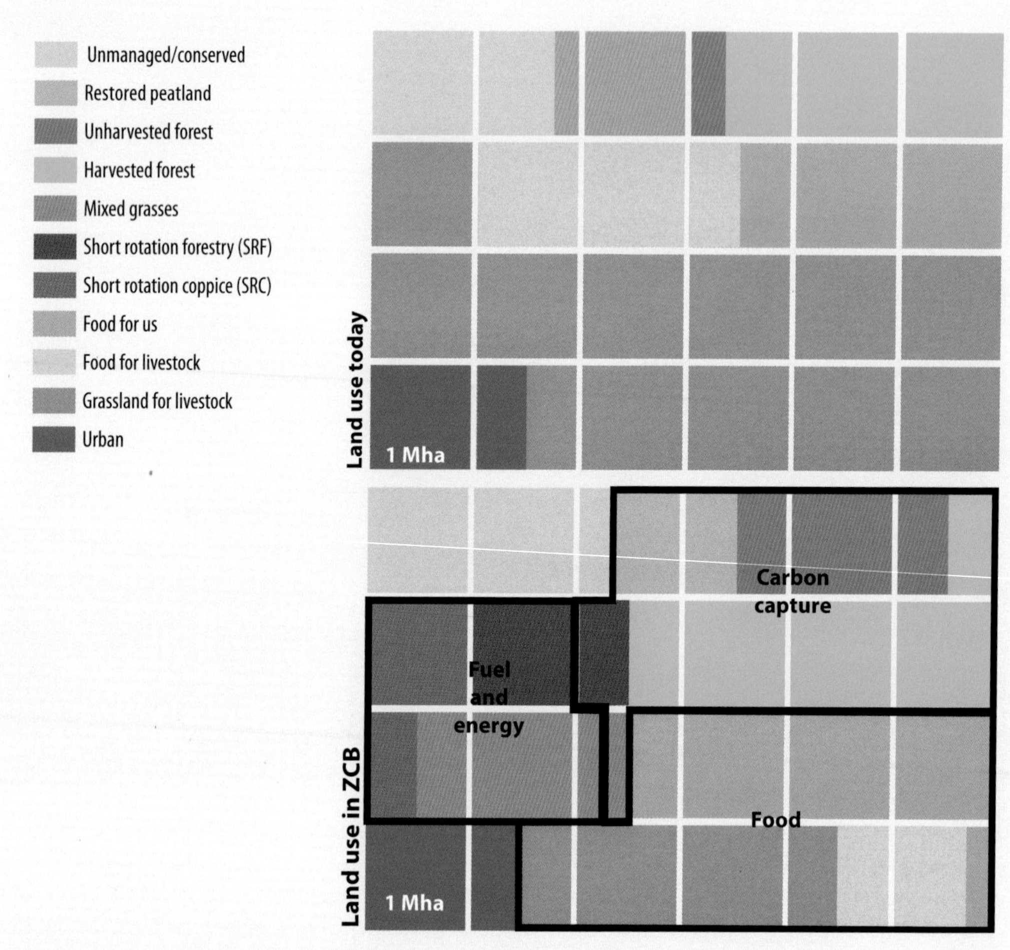

Figure 3.24: Change in land use between today (based on data from Morton et al. (2008), Forestry Commission (2007), DEFRA (2012), NERC (2008), Bain et al. (2011) and Read et al. (2009)) and our scenario. Approximate areas dedicated to providing food, fuel and energy, and carbon capture are shown in our scenario.

- Agricultural GHG emissions are reduced from 63.4 $MtCO_2e$ to about 17 $MtCO_2e$ per year via a combination of dietary changes, waste reduction, elimination of land conversion for agricultural purposes and improved land management practices.
- There is much less protein in the diet from meat and dairy sources, and more from plant sources like beans, legumes, cereals and vegetables. This results in a healthier and more balanced average diet for the UK population.
- The amount of grassland required for grazing livestock is only a quarter of the area used today (2.8 million hectares (Mha)). The same amount of cropland is used, though more of it is used to grow food for our consumption, rather than for feed for livestock.
- Whilst re-purposing this land to cater for other needs in our scenario, we take care to minimise carbon lost from soils by trying to match new land uses to the types of land 'freed up' by reducing levels of grazing livestock.

- Roughly 4.1 Mha of land (most of which was previously used for the intensive grazing of livestock) is used to produce energy by growing various grasses, Short Rotation Forestry and Coppice. In total, about 237 TWh of biomass energy is produced, adding to the 37 TWh of biomass from waste (see 3.5.2 *Waste*).
- Forest area is doubled to 24% of the land area of the UK – roughly one third of which is unharvested, and two-thirds is harvested for timber. These forests, the wood products produced and the restoration of 50% of UK peatlands, results in the capture of about 45 $MtCO_2e$ on average every year – this is required to balance the remaining emissions in the scenario and make the UK net zero carbon.
- Overall, there is more room for biodiversity in wild, conservation or protected areas.

A note on land use in our scenario

We don't break down the types of land that are used for agriculture any further than 'cropland', and 'grassland' (of three different types – temporary, intensively grazed and semi-natural grasslands). In reality, these include a wide variety of types of soil, topography (whether an area is flat or mountainous) and climate (the most northern points of Scotland in contrast to southern England). Although all of the land allocated for various uses in our scenario is currently used for agriculture, a more detailed analysis would give a better assessment of whether particular, or alternative, farming and production practices would be most appropriate, and whether or not better (or worse) yields could be expected. It would also give us the opportunity to research and incorporate organic and other farming practices, and local knowledge of an area, in our scenario, and to use this to further influence our choices.

3.6.1 Agriculture, food and diets

This section covers emissions from agricultural systems that produce food. It shows that we can reduce these emissions whilst improving the healthiness of the average UK diet. This also has implications for how we use our land, and for global agricultural systems.

Summary

- Agricultural food production is responsible for just under 10% of total UK GHG emissions – about 63.4 $MtCO_2e$ in 2010.
- The UK's agricultural GHG emissions can be dramatically reduced through changing the mix of foods in our diet: less meat, more fruit and vegetables, pulses and starchy foods (such as pasta, bread and potatoes). These proposed dietary changes would have positive health outcomes: reducing levels of obesity and diet-related diseases.
- Reducing how much beef, lamb and dairy we eat not only reduces GHG emissions significantly, but also frees up large amounts of both grassland and cropland.
- Reducing the amount of food wasted on the farm, throughout the supply chain and at home would greatly reduce food production burdens, and hence GHG emissions.
- The UK could become more self-sufficient in food production, reducing imports and the impact of food production for our consumption elsewhere in the world.
- In our scenario, emissions from food production ('on the farm') are reduced to 17 $MtCO_2e$ per year – about 27% of what they were in 2010. Imports are reduced from 42% to 17%. Land used for food production is reduced from about 78% of total UK land to about a third, freeing up space – all grassland – for other uses.

What's the problem?

The mix of different foods in our diet (meat, dairy, starchy foods, fruits and vegetables) affects the amount of land needed to feed our population, the

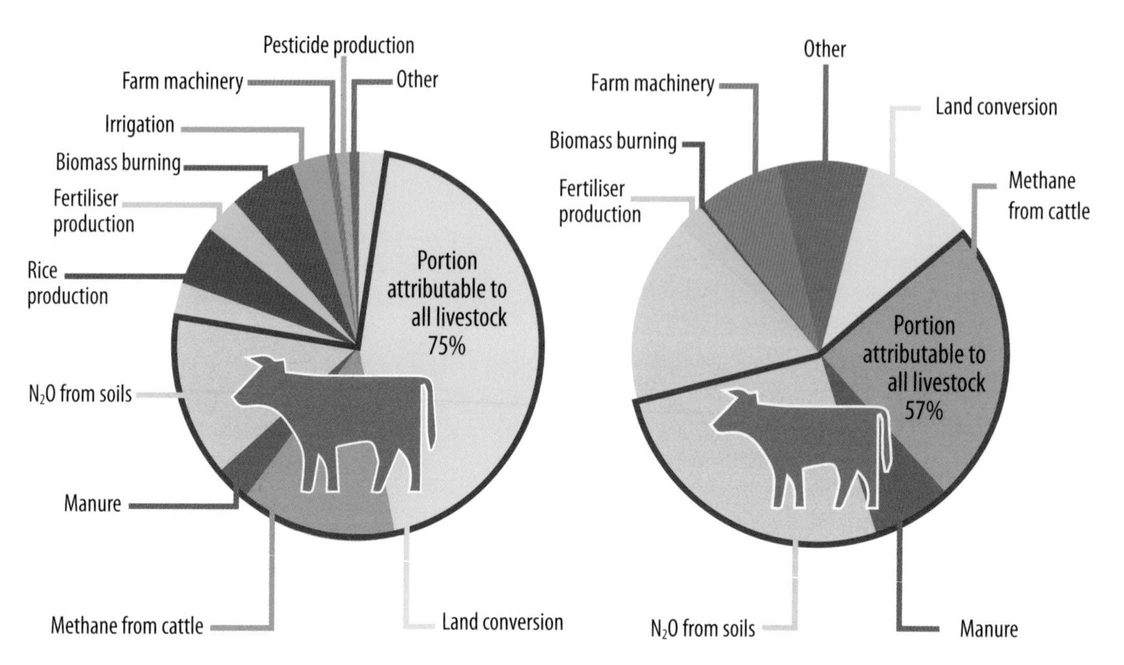

Figure 3.25: World and UK GHG emissions associated with agriculture, showing the proportions attributable to all livestock. Adapted from Garnett (2007).

GHG emissions from food production, and also has a significant impact on our health and wellbeing.

Greenhouse gas emissions

Agricultural food production is responsible for just under 10% of total UK GHG emissions – about 63.4 $MtCO_2e$ in 2010 (DECC, 2012). Figure 3.25 shows a breakdown of all emissions associated with agriculture.

Currently in the UK, about 31% of GHG emissions relating to the agricultural sector come from methane (CH_4). Nitrous oxide (N_2O) emissions, however, are the largest source of GHG emissions from UK agriculture (42%). Both methane and nitrous oxide are much more powerful GHGs than carbon dioxide (CO_2) (ibid.). The proportion of non-CO_2 emissions in agriculture is unusually high when compared to other sectors. And they can be harder to reduce, as they originate mainly from biological rather than technological sources, plus pressure is on the food system to keep producing high yields (larger amounts of food on smaller amounts of land).

Some of the CO_2 from agriculture is emitted from fossil fuel powered agricultural machinery (tractors and combine harvesters, for example – 'heavy commercial vehicles') and fertiliser manufacture. In the UK, this currently makes up about 9% of our agricultural emissions (ibid.). CO_2 is also emitted in other areas of the food supply chain (for example, in processing, packaging, distribution). All of these emissions are included in the energy and non-energy emissions from business, industry and transport (see *3.3 Power Down* and *3.5 Non-energy emissions*) and are therefore not discussed further here.

Other agricultural GHG emissions from food production that are produced 'on the farm' come from:

Converting land for food production: There are two main types of land used to produce food:

- Grassland (both intensively grazed pasture, and semi-natural grassland) for grazing livestock (mainly sheep and cows).
- Cropland (or arable land) to grow crops (for example, wheat, vegetables and sugar beet) for us to eat and to feed livestock.

Globally, the majority (47%) of emissions from agriculture still come from releasing CO_2 by converting land to agricultural use (Millstone and Lang, 2008). Plants (biomass) and soils store CO_2. Converting land for food production – either crops or livestock – releases CO_2 from the soil and the removed plant material. For example, clearing rainforest to make space to rear livestock releases CO_2 into the atmosphere (Friel et al, 2009).

The vast majority (78%) of UK land has already been cleared for agriculture. Some emissions remain from converting land to cropland (about 6.2 $MtCO_2e$ per year). Foods produced outside of the UK (the foods we import) also result in significant land use changes and are responsible for GHG emissions overseas (see box).

Meat and dairy: Cows and sheep release methane gas from their mouths as they digest grass. This process (known as 'enteric fermentation') accounts for 14% of global agricultural emissions (Millstone and Lang, 2008), but a greater 24% of the UK's agricultural emissions (DECC, 2012). UK methane emissions are higher than the global average as much more of our agricultural land is grassland (61%) dedicated to grazing livestock for meat and dairy production (DEFRA, 2011).

Other animals, like pigs and chickens, do not emit methane through enteric fermentation. They do, however, require food to be grown on cropland to feed them. In this way, they contribute to emissions from fertiliser use (see below) and can also contribute to emissions from land use change. The manure they produce is also responsible for a small amount of methane emissions (about 0.8 $MtCO_2e$ per year – 1.3% of total UK agricultural emissions in 2010) (DECC, 2012).

GHG emissions from land use change abroad

In some accounts, GHG emissions from land use change abroad attributable to food consumption in the UK amounts to as much as 100 $MtCO_2e$ per year, though our knowledge about the extent of this issue is incomplete, and different studies substantially disagree on the figure. It is a very complex issue, but it is estimated that the problem arises largely from consumption of livestock products within a globalised market – for example, clearing rainforest to rear cattle that we then import and eat, or to grow feed for UK livestock (Audsley et al., 2009). See *3.10.3 Carbon omissions* for further discussion on this point.

Rice: Methane is also released during paddy rice production, where rice is grown in fields that are flooded or irrigated. This generates about 5% of global agricultural emissions (Millstone and Lang, 2008). No methane emissions from rice production occur in the UK, however, as we do not grow rice here.

Nitrogen fertilisers: The nitrogen present in fertilisers is not taken up entirely by the crops on which they are used. Bacteria in the soil break down what remains in a process called 'denitrification'. This process releases nitrous oxide (N_2O), a powerful GHG, into the atmosphere (Di and Cameron, 2012). Direct N_2O emissions from soils occur both on land used to produce crops (both for us and to feed livestock – currently, about half of our cropland is used to grow feed for livestock (DEFRA, 2011)), and on land that is used to graze cows and sheep. Therefore, some N_2O emissions are also associated with livestock. Figure 3.25 highlights the high proportion of total agricultural GHG emissions in the UK, and globally, attributed to all livestock (cows, sheep, pigs, chickens etc.).

What we eat

Our diets supply the energy we require for daily activities and basic metabolic processes (like breathing and keeping our heart beating).

The food we eat also provides essential amino acids (from protein), vitamins and minerals, essential fatty acids (such as omega 3) and anti-oxidants that help prevent disease. Eating an unhealthy diet for a long time can lead to many diet-related diseases like heart disease and diabetes (Friel et al, 2009).

In the UK, *on average* (we realise that we are all individuals, and in some cases these issues are irrelevant!), we currently eat:

Too much food: Eating too much food can make a person overweight or obese, and at greater risk of specific diet-related diseases, such as heart and circulatory problems (heart disease and high blood pressure, for example), strokes, type II diabetes, and certain cancers (like colorectal) (ibid.). In the UK today, 64% of adults are overweight or obese (Bates et al, 2011), and 71% of all deaths in the UK in 2010 were from the types of diseases mentioned here (WHO, 2013). Physical activity levels are also decreasing. Increased car use, office jobs and increased television viewing are some of the reasons given for this rise in sedentary lifestyles (Poskitt, 2009).

Another problem for the agricultural industry is that a lot of food produced is also wasted: both throughout the food chain and within the household. Just over 30% of all food produced in Europe is wasted (FAO, 2011). This means that with overconsumption and wastage, levels of production are a lot higher than they need to be.

An unhealthy balance of foods: Many developed countries are eating diets that are becoming less and less 'balanced' – too much of some foods and not enough of others. Sweets, crisps and cakes provide us with lots of energy (measured in kilocalories (kcal)) but very few nutrients. Fruits and vegetables, on the other hand, are high in nutrients and relatively low in energy (Monsivais and Drewnowski, 2007).

An average UK citizen today eats 2,630 kcal in energy terms, and 80 grams (g) of protein per day. Both are too high – about 2,250 kcal and 55 g respectively are recommended daily amounts (RDA) (COMA, 1991; FSA, 2007). The average UK diet also doesn't meet fruit and vegetable recommendations (of at least five portions a day), or the recommendations for cereal and fibre intake. Our average diet contains too many foods that are high in fat (particularly saturated fat), salt and sugar (known as 'high in fat, salt and sugar (HFSS)' foods), and we eat too much red and processed meat. This shows that there is a problem with the current *mix* of foods within our diet as well as with overall calorie consumption (ibid.).

What's the solution?

Many solutions reduce GHG emissions and address health issues together. For example, for us in the UK, eating less red meat should be recommended from both a GHG emissions perspective and a health perspective. These changes would also reduce land use change, and enable emissions reductions overseas (McMichael et al, 2007).

Greenhouse gas emissions

There are many ways we can reduce emissions from food production:

Minimising land for food production, and managing it better: Reducing CO_2 emissions from agricultural land use, both at home and overseas could be achieved by minimising the amount of land converted to agricultural production, or stopping agricultural expansion completely (especially into forests, peatland and less intensively managed, or semi-natural landscapes). There are

Agricultural emissions overseas

Emissions from the UK food chain amount to 115 $MtCO_2e$ (this includes transportation and processing of goods – energy emissions included in *3.3 Power Down* and *3.4 Power Up*). Because we import 42% of all the food we eat it is likely we are responsible for a great deal more emissions globally – at least a further 59 $MtCO_2e$, not including land use change (Holding et al., 2011). Emissions relating to imports are not included in the 'production' GHG emissions accounting system as these emissions do not occur on UK territory. If a consumption based accounting system is used instead, however, overseas emissions relating to imports would be included (see *3.10.3 Carbon omissions*).

What about 'GM' crops?

Crops can be genetically modified (GM) to produce more food using less land – in other words, higher yields. Whether or not to grow GM crops is currently hotly debated. In Europe the use of GM crops is restricted, but in America they are already grown widely (Devos et al, 2009).

Some people believe that the use of GM crops is inevitable because we will need more food to feed growing populations (Godfray et al., 2010). Others question this premise and have concerns over their long-term sustainability and the legal ownership of food (War on Want, 2012).

Because of current concerns, we do not use GM crops in this scenario. However, *if* they were proved safe and managed correctly, higher yields could be useful if the UK population continues to grow.

also management techniques we can use to encourage agricultural soils to instead capture carbon (see *3.6.3: Capturing carbon*).

We could even use *less* land to produce food, and restore some of it to more natural landscapes – adding carbon to soils in some cases, rather than releasing it (again, see *3.6.3: Capturing carbon*). There are a number of ways of doing this, even with a growing population:

- **Product switch:** change the mix of foods we eat. This means eating fewer land-intensive foods, such as those from grazing livestock (beef, lamb and dairy), and replacing them with types of food that require less land – see figure 3.26 for examples of the land use implications of some dietary choices.
- **Intensify production:** grow more intensively on the land we do have (higher yields). This could, however, compromise soil quality and the amount of carbon it holds. It may also require more fertiliser, increasing the emissions from their production and use, and more pesticides, which may have a negative impact on biodiversity (Miraglia et al., 2009). We could use glasshouses for food production, allowing us to grow more crops at higher yields. Renewable energies and waste heat could be used to provide the heat and light that glasshouses need; alternatively we could use geothermal heat (Sinclair Knight Merz, 2012).
- **Increase imports:** importing more food from abroad would mean we use less land in the UK. Although this might be good for our GHG emissions, it would also mean increasing reliance and stress on land elsewhere, and potentially increasing emissions from land use change and agriculture overseas (see boxes on pages 85 and 86).

Less meat (particularly red meat) and dairy: 'Technical fixes' for reducing methane emissions from cows and sheep are not yet fully tested – see boxes on pages 88 and 89 (Eckard et al.). Currently, the only feasible way of reducing methane emissions from grazing livestock is to reduce the number of cows and sheep. Since over 11 Mha of grassland is currently used for grazing livestock in the UK – both intensively grazed and semi-natural grassland – reducing the number of grazing livestock also has the advantage of freeing up substantial portions of land for other uses (DEFRA, 2011).

Plant-based sources of protein require much less land and emit far fewer GHGs than animal-based proteins, even if larger amounts (by weight) have to be eaten to get recommended amounts of protein. Figure 3.26 shows four high protein food sources and compares the land needed to produce the amount of that food required to satisfy a recommended daily allowance (RDA) of protein, and the associated agricultural GHG emissions (those produced 'on the farm', not including fossil fuel use). It suggests that changing the mix of different high protein foods in the diet – from animal-based to plant-based sources – can result in large GHG emissions and land use reductions without having to compromise on nutritional health. Protein deficiency in most cases is not associated with a lack of meat, but with not enough or a poor variety of other foods (Gonzalez et al.).

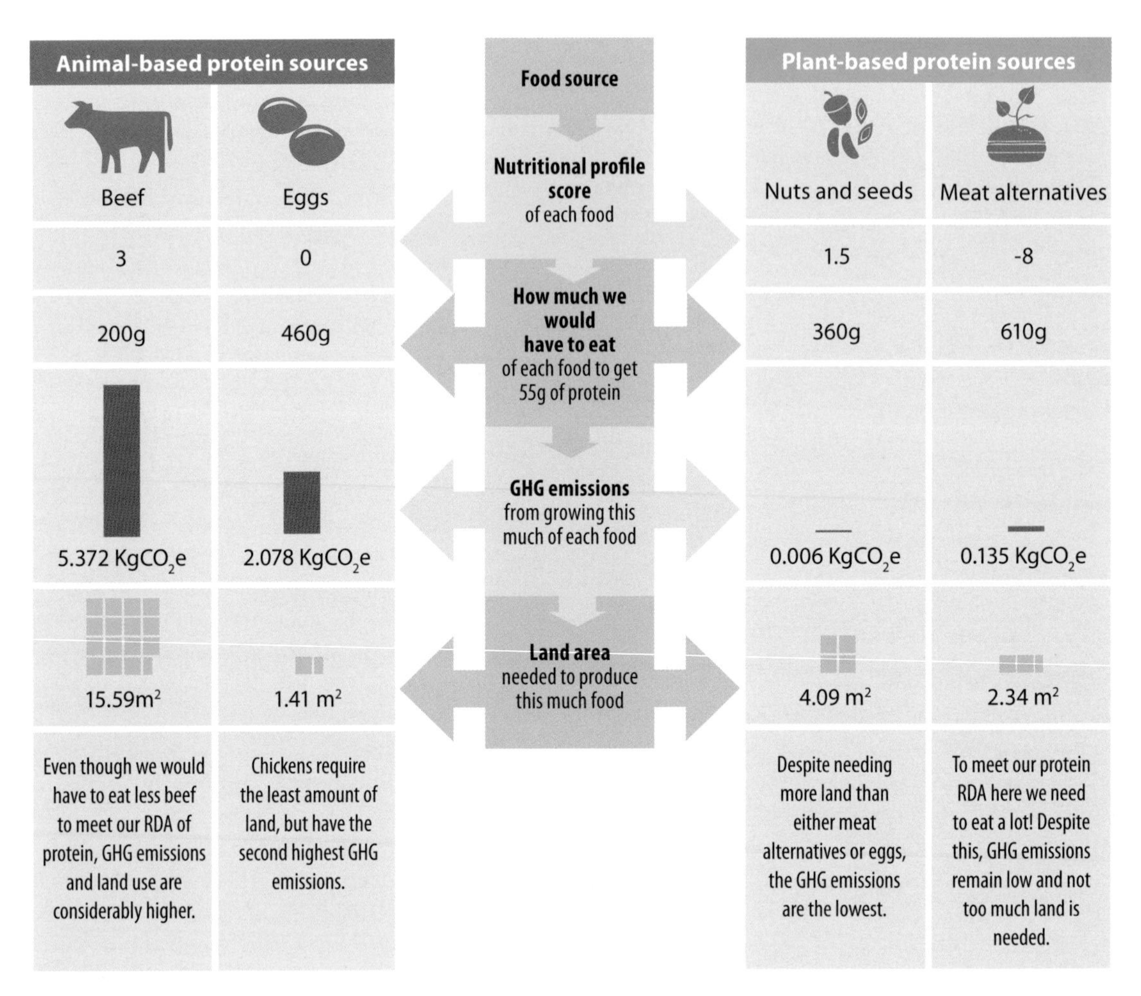

Figure 3.26: Comparison of four different high protein food sources: their Nutritional Profile Scores (NPS), how much would need to be eaten to meet the recommended daily amount (RDA) of protein, and the associated agricultural GHG emissions and land used for producing this amount of each food source.

Are there other ways of reducing methane emissions from grazing livestock?

The methods below have not been fully proven or tested, and thus require more research (Eckard et al., 2010). For this reason they are not included in our scenario, but could prove useful in the future, should they be proven successful. They are:

- **Selective breeding.** We may be able to choose animals that naturally emit lower levels of methane, reducing emissions by 10-20% (Hongmin et al., 2011). However, it is currently very expensive to test for these traits and very difficult to do on a large scale.
- **Vaccinations.** These are being developed to block the processes within the digestive system that produce methane, although significant emissions reductions have not yet been found (Buddle et al., 2010).
- **Changes in the diet of grazing livestock.** An increase in fatty acid intake is being studied as a potential way of reducing associated methane emissions. Emission reductions in this study, however, only averaged at 3% (Grainger and Beauchemin, 2011).

What about replacing meat with 'cultured meat'?

The benefits of this type of meat production would be its incredibly low land and water usage, as well as its significant reduction in GHG emissions (Tuomisto and de Mattos, 2010). Two techniques are currently being trialled:

- **Lab produced meat from stem cells.** It is already possible to grow basic muscle and fat tissue (the main parts of an animal that we eat, aside from offal) in this way, but the process doesn't successfully mimic the taste and texture of meat (Post, 2012). The different types of tissue also have to be grown individually, meaning what is grown is only really suitable for ground meat, which is used, for example, in burgers (Datar and Betti, 2009).
- **Organ printing.** This technique is currently being developed within medical research to make human organs for transplants. Live cells are sprayed onto gels in layers (in the same way that inks are printed onto paper) to make 3D structures. This method could provide more realistic taste and texture, and even produce individual cuts of meat (such as a steak or lamb chop) (Mironov et al., 2009).

Both of these technologies are still at the research stage and are not currently viable for mass production (Bhat and Bhat, 2011), but perhaps offer an opportunity for the future.

Different rice: We can reduce methane emissions from rice production by importing more of our rice from rice crops not grown in paddy fields. These alternative methods of production can reduce field GHG emissions from rice production by up to 50% (Blengini and Busto, 2009).

Nitrogen inhibitors: Nitrogen inhibitors' (NIs) can be mixed into fertilisers – they block the conversion to N_2O (nitrogen oxide, a GHG) so that more nitrogen remains in the soil and less N_2O is emitted. This can also improve yields (plants take up more nitrogen and so grow better) and could result in less fertiliser being required (Di and Cameron, 2012). How effective NIs are is dependent on many factors. Studies have shown average reductions in N_2O emissions of between 38% and 49% (Liu et al., 2013; Akiyama et al., 2010 respectively).

Using NIs on land used to grow feed or graze cows and sheep also lowers the associated N_2O emissions (Liu et al., 2013).

What we should eat

On average (again, these rules don't apply to each individual!) for a healthy diet, we in the UK could:

Eat less food: We (as a nation) need to rebalance our energy levels (eat the right amount of kilocalories for daily activities and basic metabolic processes). We can do this by either eating less or becoming more physically active, or (most effectively) a combination of both. This will help lower the incidence of diet-related diseases seen in the UK today (Lang and Rayner, 2007). 'Becoming more physically active' can result from efforts to reduce GHG emissions from transport, like walking and cycling more (see 3.3.2 *Transport*).

A better balance of foods: No food need be completely off limits, but foods do vary significantly in their nutritional qualities. Nutritional profile scores (NPS) have been developed in order to highlight these variations and demonstrate clearly a food's nutritional merits or failings. These scores make it easier to tell at a glance how good or bad a food is and how it compares to other foods (Stockley et al., 2007). Figure 3.26 shows some examples of the NPS of four protein sources – lower or negative values are best.

The government also offers advice on good food balance (the right proportions of different types of foods) in the form of an 'Eatwell Plate'. Based on these guidelines, we developed a number of criteria to assess a diet, both 'essential' and 'ideal'. The essential criteria relate to things that have been *proven* to promote health and lower disease risk (see WHO, 2003 and Pan et al., 2012 for two examples). The ideal criteria are simply *recommended* for a healthy diet (FSA, 2007).

Essential criteria:

- A minimum of five portions of fruit and vegetables per day.
- About a third of the diet made up of starchy foods (for example, pasta, rice, bread and potatoes (not fried)).
- No more than 10% of daily energy intake (kcal) made up of unhealthy foods high in fats, sugar and salt (HFSS).
- No more than 70 g of red and processed meats eaten per day.

Ideal criteria:

- Wholegrain cereals (such as brown rice and bread) chosen where possible.
- More plant-based protein, such as pulses (lentils, chickpeas and baked beans). These are much lower in saturated fats than animal-based protein.
- More 'good fats' from foods like oily fish, nuts, seeds and vegetable oils, rather than 'bad fats' from foods like butter, cheese, crisps, sweets, biscuits, cakes and chocolate.
- Less battered and fried chicken than other forms of chicken.
- Skimmed and semi-skimmed milk chosen over whole milk.

Our scenario

It is completely feasible for the UK population to have an average diet that is both lower in GHG emissions and healthier.

In our scenario we become more self-sufficient, importing only 17% of our food products rather than the current 42%. Most importantly, we do not import livestock products or feed for livestock (see *3.1 About our scenario*). This has the additional benefit of reducing demand for land in other countries, thus helping to prevent emissions from agricultural land use change overseas (though it can be difficult to exactly quantify the effect – see *3.10.3 Carbon omissions*).

Furthermore, in our scenario the UK can provide a healthy diet for a growing population not only without converting new land to agriculture but by actually reducing the amount of agricultural land needed in the UK. This has positive consequences for energy and fuel production, and for the protection and conservation of natural landscapes, as well as for GHG emissions reductions (see *3.6.2 Growing energy and food* and *3.6.3 Capturing carbon*).

What does this average diet look like?

Again, it is important to remember that the average diet is not prescriptive – that is, to be followed exactly by everyone. Averages do not reflect the differences between recommendations for men and women, or between those of different age groups. Neither do they reflect the wide range of personal preferences or cultural choices. An average does, however, provide an idea of what might change in the consumption patterns of a population.

In our scenario, every person, on average, eats (in energy terms) about 2,280 kcal per day. Most of the energy we need comes from starchy foods like pasta, rice and potatoes; unhealthy foods, such as cakes, biscuits and crisps (HFSS foods) are eaten much less regularly than today.

On average, each person's daily protein needs come from a weekly combination of:

- One large portion of red meat per week (a steak, pork or lamb chop, a portion of liver or a chicken fillet).
- Two further smaller portions of pig or chicken per week (for example, two rashers of bacon or a few slices of chicken).
- A fillet of fish.
- One portion of meat alternatives, such as tofu.
- Four portions of pulses (such as lentils, chickpeas and baked beans).
- Two eggs.
- Enough milk to cover breakfast cereal and cups of tea and coffee (with additional milk coming from alternatives, such as soya).
- A small portion of cheese and yoghurt.

The amount of protein this combination supplies (along with proteins in other foods such as cereals

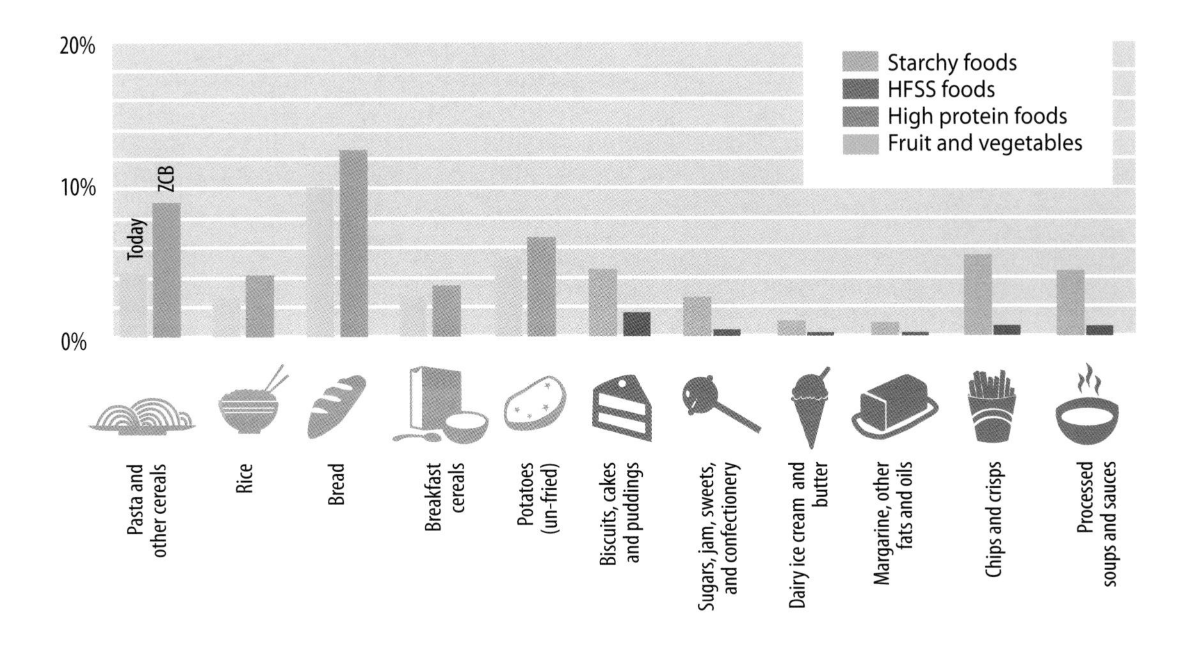

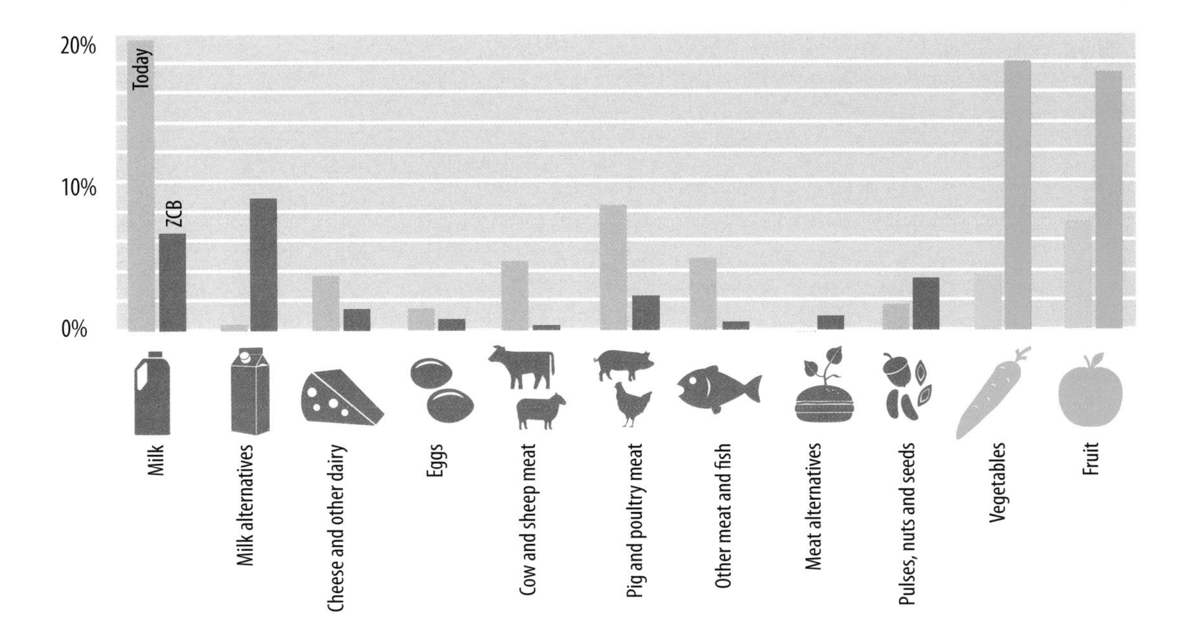

Figure 3.27: Percentage contributions of different types of foods in the diet (Bates et al., 2012) The difference between the average diet today and in our scenario is shown.

What about specific dietary needs?

Although our modelling looks at the broad nutritional adequacy of diets, nutrition is a very complex area of research. To get a better picture we would need to look at the provision of the whole range of nutritional needs, micronutrients for example (especially vitamin B12 and iron). It would also be good to look at the individual needs of various population sub-groups – we know that the elderly and children have different dietary requirements, for instance. Our scenario greatly improves the 'healthiness' of the average UK diet, and it is likely that individual needs could be catered for within the range available, but we have not specifically tested for this.

and vegetables) provides an average intake of 72 g, which is still higher than the RDA of 55 g. The fact that our modelled values for protein still exceed the RDA demonstrates that, on average, protein insufficiency is unlikely to occur in our scenario.

There are almost four portions of vegetables per day (one portion is 80 g) and three portions of fruit – four times more than we eat today. Figure 3.27 shows how much of our diet is made up by each of the different food categories compared to the average UK diet now.

What about removing meat from the diet altogether?

It is important to note that by far the easiest possible way of reducing emissions from food production would be for the entire population to eat no red meat and no dairy, thus eliminating methane emissions from grazing livestock and some N_2O emissions from soils. According to the latest statistics of UK diets, however, only 2% of the UK population is reported to be vegetarian or vegan (Bates et al., 2011). In the design of a new average UK diet, therefore, we attempt to balance nutritional requirements, land use restrictions and GHG emissions reductions with current taste preferences.

What impact does this have on GHG emissions?

Agricultural emissions from food production in our scenario are reduced to 17 $MtCO_2e$ per year – a 73% reduction. This represents only agricultural emissions produced 'on the farm'. Emissions from food processing and distribution are energy related emissions and so are taken into account in 3.3 *Power Down* and 3.4 *Power Up*.

The emissions reductions come from:

- Applying nitrogen inhibitors to the soil, reducing field N_2O emissions by 38%.
- Reducing total food production, even though the population is expected to increase. The amount of food produced for each person over a year is reduced from 1.1 tonnes to 0.9 tonnes per person per year. This is mainly because half the current level of food waste is assumed, and each person eats only the amount of food that is recommended, thus reducing how much food we need to produce.
- Reducing the amount of beef and lamb products in our diet by 92%.
- Pig and chicken products (including eggs) are reduced by 58%.
- Dairy consumption is also reduced; products such as milk, cheese and yoghurt are reduced by 59%.

It is also worth noting here that agricultural GHG emissions from sugar are some of the lowest amongst all crops grown in the UK. If we only consider GHG emissions, we could eat a lot more sugar crops. Restrictions on these products in our scenario are for health and land use reasons.

What impact does it have on land use?

Figure 3.28 shows how the land we use for food production changes from the current UK situation to that in our scenario. In our scenario about 7.4 Mha of UK land is used for food production. The area of cropland required (4.6 Mha) is about the same as today. The demand for feed is reduced in our

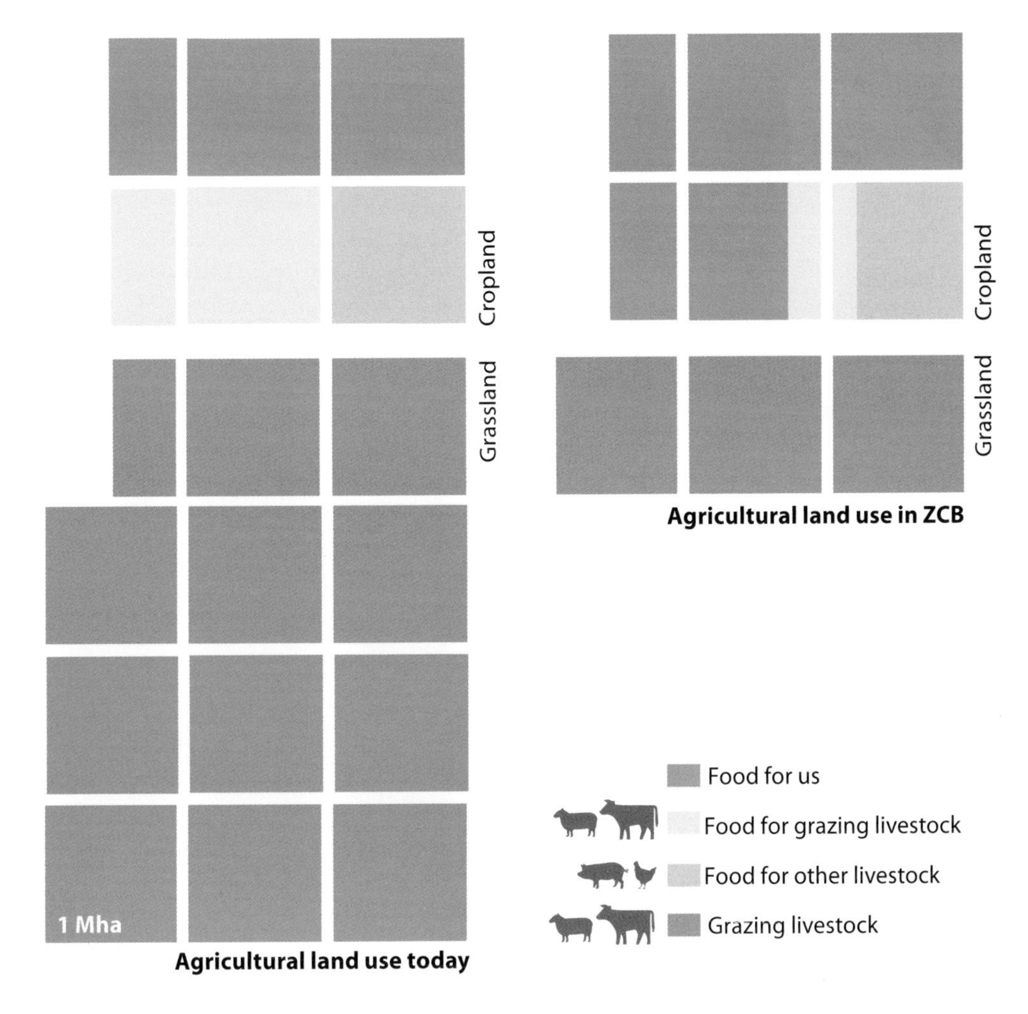

Figure 3.28: The area of cropland and grassland used for agriculture today (DEFRA, 2012) and in our scenario.

scenario; only about a quarter of our cropland is now used to grow feed. The rest of the cropland (74%) is used to grow food for us to eat. Of this:

- 1.6 Mha is used to grow starchy foods (cereals).
- 1.3 Mha is used to grow fruit and vegetables. Fruit and vegetable production increases fourfold. This is because we import less and produce more at home, and because how much we eat increases. The number of hectares used for glasshouses also doubles. This enables us to grow more salad vegetables (such as tomatoes, cucumbers and peppers), as well as some of our own pulses (such as lentils and kidney beans).
- 0.2 Mha for HFSS foods. As we eat less HFSS foods, the *total* amount of land dedicated to sugar and oils is reduced dramatically. As we can grow oil crops and sugar beet in the UK, however, all sugar and oil production is brought home meaning that UK land dedicated to these products actually increases slightly.
- A further 0.3 Mha is used for other foods – nuts, seeds and 'meat alternatives' for example.

The remaining imports are grown on cropland overseas – an estimated 1.2 Mha abroad is used to

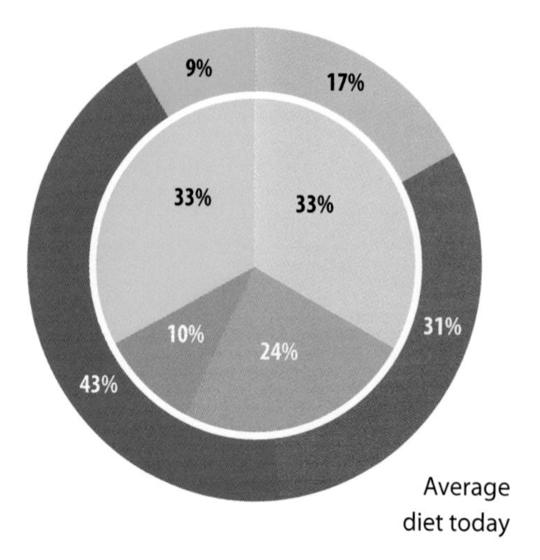

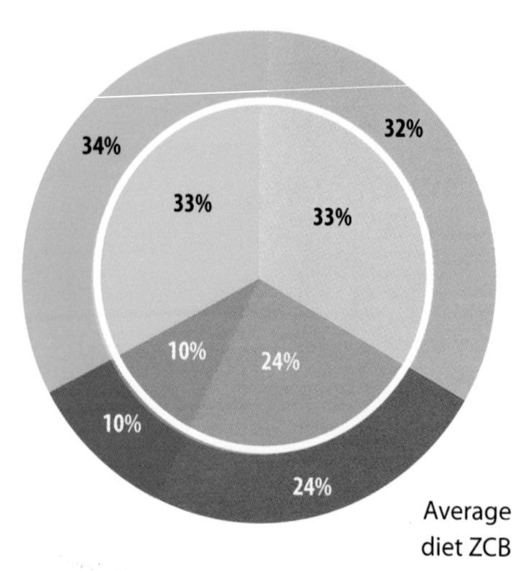

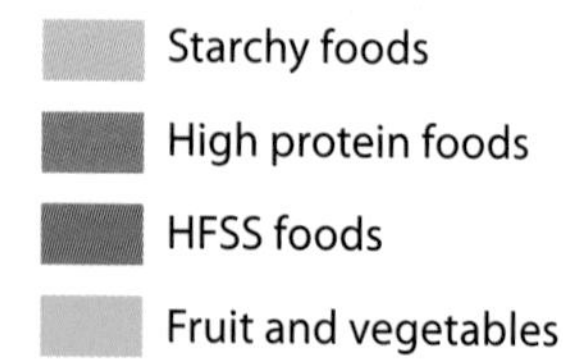

Figure 3.29: The Eatwell Plate. Government recommendations for a healthy balanced diet (FSA, 2007). Today's average diet and the average diet in our scenario are shown (outside circle) relative to the Eatwell Plate recommendations (central circle).

grow cocoa beans, rice and tropical fruits – things we cannot grow in the UK.

The amount of grassland required to graze livestock is only a quarter of the area used today (2.8 Mha) – some of which is intensively grazed grassland, and some of which remains as semi-natural grassland. The amount of grassland for meat production is reduced by 82% and the amount of land for dairy cows is reduced by 65%.

What impact does this have on the UK's health?

Since our proposed average diet is specifically designed to have a positive impact on the UK's health, this question is easy to answer: the suggested average diet in our scenario both satisfies health recommendations and meets nutritional requirements. The average diet in our scenario meets government recommendations for a healthy food balance (see figure 3.29).

In the average diet in our scenario all four 'essential' dietary balance criteria are met, in contrast to today's average diet where none are covered. All five 'ideal' criteria are also met, whereas today's average diet meets only two (Bates et al., 2011). The dietary changes in our scenario are in line with nutritional recommendations for lowering levels of obesity and diet-related diseases, and so also improve the health of the UK population in this way.

3.6.2 Growing energy and fuel

This section shows how we can grow various energy crops in the UK to provide biomass to cover energy demands that cannot be met with electricity.

Summary

- In our scenario, industry and transport require biomass for heat, and synthetic liquid and gas. Back up for our renewable energy supply also requires synthetic gas. This energy demand comes to 234 TWh per year (see 3.3 *Power Down*).
- Around 274 TWh of biomass is required every year to meet these demands (see 3.4 *Power Up*). 37 TWh

of this biomass comes from waste (see 3.5.2 *Waste*). The remaining 237 TWh comes from specifically grown energy crops.

- 4.1 Mha of land is converted to growing energy crops, most of which is currently used to graze livestock.
- Second generation energy crops grown on this land with low inputs and without significant release of carbon from soils are Short Rotation Forestry (SRF), Short Rotation Coppice (SRC), Miscanthus (also known as 'elephant grass') and other mixed grasses.
- The yields of Short Rotation Coppice, Miscanthus and other grasses are expected to increase in the future, which helps produce more biomass from less land.

What's the problem?

Sections 3.3 *Power Down* and 3.4 *Power Up* show that some energy demands cannot be met by electricity. Some demands require fuels with high energy densities by weight and by volume – ones that are easily stored and transported because they are small and light. Fossil fuels are currently particularly useful in these cases. Alternatives that do not emit GHGs are required to provide all our energy requirements with zero carbon emissions.

In total, some 234 TWh per year of this type of energy is required in our scenario. These energy demands are:

- **Buildings and industry:** (*3.3.1*) 36 TWh of biomass for heat per year (10 TWh for buildings, and 26 TWh for some industrial processes); 61 TWh of biogas or synthetic liquid gas, and 12 TWh of synthetic liquid fuels, also for industrial processes.
- **Transport:** (*3.3.2*) 98 TWh per year of synthetic liquid fuel (39 TWh for aviation, 59 TWh for heavy commercial vehicles and ships).
- **Balancing supply and demand:** (*3.4.2*) about 27 TWh per year of biogas or synthetic gas as back up for our energy supply.

This means we need to produce 110 TWh of synthetic liquid fuel, 36 TWh of biomass for heat, and 88 TWh of synthetic gas or biogas, but without emitting GHGs.

What's the solution?

Carbon neutral fuel replacements

Biomass from energy crops can be used to make fuels with identical (or sufficiently similar) characteristics to fossil fuels. However, we cannot solve the entire energy problem by growing biomass – there is simply not enough land. In fact, to provide enough biomass to satisfy *all* our energy demands today we would need an area at least twice the size of Britain.

That said, the use of some biomass is essential because it provides storable energy, and can provide gaseous and liquid fuels through various chemical processes. Though biofuel and biogas can be created from biomass directly, *3.4.2 Balancing supply and demand* and 3.4.3 *Transport and industrial fuels* show that biomass can be combined with hydrogen to produce synthetic gas and liquid fuels, which increases the amount of fuel produced per unit of land. Though there are significant losses in these processes, the hydrogen required can be made using surplus electricity (at times of high supply and low demand), meaning we do not have to have additional infrastructure to produce it. These chemical processes are called the Fischer-Tropsch (FT) process (which produces synthetic liquid fuels), and the Sabatier process (which produces synthetic gas).

These synthetic gas and liquid fuels are 'carbon neutral'. The CO_2 emitted by burning them was initially taken in by the biomass as it grew, and the electricity used is renewably produced. Over the long-term there is no net increase of GHG emissions in the atmosphere.

Energy crops in the UK

Various types of energy crops are suitable for growing in the UK. They vary in how much biomass they produce (the 'yield'), and how suitable they are for various land types. Some energy crops need very good quality cropland to grow on, which is limited and usually already used for growing food. For example, **first generation biofuels** are made from wheat, corn, sugar crops and vegetable oil – all of which could alternatively be eaten. **Second generation biofuels**, in contrast, are from 'woody' plant material and non-food grasses, which can be grown on grassland that is currently used for grazing livestock.

The yields of some energy crops are expected to increase by up to 100% in the future (DEFRA, 2009). This will be achieved through plant breeding, to produce and improve strains specifically for energy production. However, the emphasis must be on gaining high yields with low fertiliser and water requirements (Sims, 2006).

The main **second generation** energy crops are as follows (Biomass Energy Centre, 2011):

Short Rotation Forestry: Short Rotation Forestry (SRF) is the closest energy crop to conventional forestry and uses fast-growing native tree species, such as birch, alder and sycamore. These species grow well on many different qualities of land. These trees grow much faster than many conventional timber-producing species – SRF is usually cut back after 2-4 years, or felled after 8-20 years of growth, and then replanted. However, this is a much slower 'turnaround' than many other energy crops, and yields generally aren't as high.

The biomass produced from SRF can be burnt directly to produce heat, or in Combined Heat and Power (CHP) systems.

Short Rotation Coppice: Short Rotation Coppice (SRC) is usually made up of willows and poplars, which are 'coppiced' after a few years. The main woody material of these plants is harvested, but the roots remain and will regrow – it is a perennial plant. The whole coppice needs replanting only every 30 years or so. It grows well on various different qualities of land, and yields are expected to increase in the future.

Biomass from SRC is very flexible in its use – it can be burned directly for heat, used to make biofuel or biogas directly, or to produce synthetic biogas or synthetic liquid fuels.

Miscanthus: Miscanthus is a tall grass – sometimes known as 'elephant grass' – that is harvested every year to grow back the following year (also a perennial crop). As a dedicated energy crop, Miscanthus has high yields, which are expected to increase substantially in the future.

Miscanthus can also be burned to provide heat,

and can be used to make biogas or biofuel directly, or to produce synthetic gas or liquid fuels.

Other grasses: Other grasses can also be used for energy production. They are harvested 'green' (with a high moisture content) and are best used to produce biogas through anaerobic digestion (AD). They can be grown on various different land types, and the most suitable species can be chosen depending on local conditions. Growing mixed species can help improve the biodiversity of the area.

Our scenario

In our scenario we try to match the needs of our energy system with the needs of our land – that is, we try to match the energy crops to the most suitable types of land that are 'freed up' when we reduce the amount of grazing livestock. This helps minimise carbon lost from soils, which can occur when we change the way we use our land (see *3.6.3 Capturing carbon*), but also limits the amount of land we can use to grow energy and fuel.

Some land currently used as temporary agricultural grassland (around 0.9 Mha) continues as such, but with mixed grasses grown as an energy crop. Most of the land made available to grow energy crops is currently intensively grazed grassland (about 2.7 Mha). This good quality grassland that is in many cases already fertilised, is used to obtain high yields of Miscanthus and SRC. Some of this land is also used to grow SRF, together with a small amount of semi-natural grassland (around 0.5 Mha), which also becomes available because it is no longer grazed.

We cautiously assume that the yields of grasses, Miscanthus and SRC increase by 50% from average yields today because of improvements through plant breeding. SRF is expected to produce similar yields to those today.

Figure 3.30 summarises the area required for these different crops, how much biomass is produced and what this biomass is used for in our scenario. Biomass from mixed grasses are used to produce carbon neutral synthetic gas and biogas by anaerobic digestion; biomass from Miscanthus and some SRC is used to produce carbon neutral synthetic liquid fuel; and the remaining biomass from Short Rotation Coppice and Short Rotation Forestry biomass is used for heat in buildings and industry.

Losses in the conversion processes from biomass to synthetic gas and liquid fuels mean that a total supply of 274 TWh per year is required to meet the 234 TWh of demand (see *3.3 Power Down*). The annual yield of all the energy crops in our scenario is about 237 TWh per year. In addition, the equivalent of 37 TWh of biomass is produced every year from sewage, manure and agricultural and crop residues – straw from cereals for example (see *3.5.2 Waste*). This is used to produce biogas through anaerobic digestion (AD). The residue from AD is reapplied to soils to recycle the nutrients and decrease the amount of fertiliser required.

In total, the biomass produced (from energy crops

What are the effects of growing energy crops?

In general, the difference between energy crops and food crops is that energy crops are grown to maximise the amount of carbohydrate, whereas food crops are more often grown to contain high levels of protein and other nutritional factors. This makes a difference to the amount of fertiliser required – crops produced specifically for energy can be bred to minimise fertiliser requirements (Sims, 2006).

A crop harvest can also be timed so that nutrients in the plants return to the soil and, in the main, only plant material containing carbohydrate is removed. By-products from burning biomass and the residue left after producing biogas by anaerobic digestion can also be returned to the soil to recycle the nutrients, further decreasing fertiliser requirements, and reducing GHG emissions from soils (see *3.6.1 Agriculture, food and diets* for information on GHG emissions due to nitrogen fertiliser use).

It is important that proposed changes to land use are assessed for their impact on wildlife. Miscanthus and Short Rotation Coppice can provide good habitats for wildlife in comparison with cropland (Haughton, 2009). However, land used for growing energy crops must be managed responsibly in order to promote biodiversity and regulate water use.

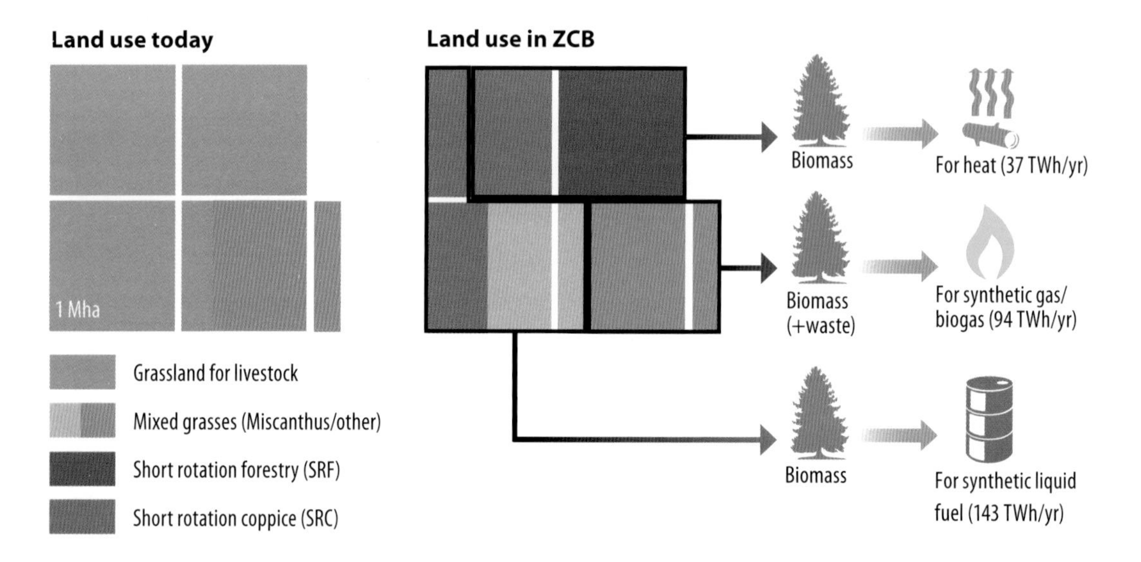

Figure 3.30: Area of land used today (DEFRA, 2012) that is used for energy crops in our scenario, the types of crop grown, and the amount and use of the biomass produced.

and waste) is used to supply energy in the following forms, to cover the various demands:

- 143 TWh per year of biomass for the production of synthetic liquid fuels.
- 94 TWh per year of biomass for the production of biogas and synthetic gas.
- 37 TWh per year of biomass for heat.

3.6.3 Capturing carbon

This section describes how we can use and manage land to reduce associated emissions, and increase the amount of carbon we capture. It shows that the total potential is limited, but that we can balance the remaining GHG emissions in our scenario.

Summary

- In our scenario, the remaining effect we have on climate change is equivalent to about 45 $MtCO_2e$ per year, despite emissions reductions of about 94% from 2010.
- By doubling the forested area of the UK, harvesting more timber to use in buildings and infrastructure, restoring 50% of our peatlands, and converting waste wood either into biochar or leaving it in 'silo stores', we capture the required 45 $MtCO_2e$ per year (on average), making our scenario net zero carbon.
- It is possible that planting more forest, or restoring more peatland could capture more carbon, though the land available means there are limits to these measures. We must be careful not to release carbon from soils in the process of land use change.
- These methods should last long enough (about 100 years) for us to develop new technologies or ways of doing things that replace the activities in our scenario that still emit GHGs.

What's the problem?

Despite GHG emissions reductions of about 94%, our scenario still has an impact on climate change equivalent to 45.2 $MtCO_2e$ per year – emissions of 15.9 $MtCO_2e$ from the non-energy emissions from industry, businesses and households; 5.1 $MtCO_2e$ from waste management; 17 $MtCO_2e$ from agriculture; plus the additional impact of flying equivalent to 7.2 $MtCO_2e$.

To become net zero carbon, we must balance this impact by capturing carbon every year.

Capturing carbon today

Carbon is being captured in the UK already:

1) New forests and grasslands take carbon into soils, trees and grass as they grow (Broadmeadow and Matthews, 2003).
2) Existing forests in the UK cover about 2.9 Mha (12%) of UK land. They were capturing over 10 $MtCO_2e$ per year in 2010 (Read et al., 2009).
3) Harvested wood (timber) stores carbon when used in construction – for example, in timber framed buildings (ibid.).

In 2010, a total of 23.7 $MtCO_2e$ was captured (see figure 3.31), according to UK GHG accounts (DECC, 2013). However, this is less than 4% of the UK's total GHG emissions, and the current carbon capture processes will not last forever:

1) Relatively little new forest has been planted over recent decades (Atkinson and Townsend, 2011).
2) As existing forests mature, they will capture less carbon year-on-year – these carbon 'stores' will eventually fill up (Smith, 2010). By 2020, only about 4.6 $MtCO_2e$ will be captured each year (Read et al., 2009).
3) The majority of British conifer forests are due for felling in the next 10-20 years (ibid.). UK timber supplies, which store carbon in wood products, are projected to decrease (Forestry Commission, 2010).

Furthermore, 19.8 $MtCO_2e$ was emitted from parts of the UK landscape in the same year. Therefore, only 3.9 $MtCO_2e$ was captured on balance (DECC, 2013) – see figure 3.31. Our management of the UK landscape is contributing to the problem and, without changes, will continue to do so:

- CO_2 was emitted from soils and plants by urban expansion into forest and grassland environments (DECC, 2013). These emissions are discussed in 3.5 *Non-energy emissions*.
- Conversion of forest and grassland to cropland, and the management of all agricultural land (both cropland and grassland) also contributed to GHG emissions in these areas. These emissions are discussed in 3.6.1 *Agriculture, food and diets*.
- UK peatlands, including (though not exclusively) 'wetlands' in figure 3.31 (some peatlands form part of cropland and grassland habitats) currently emit almost 3.7 $MtCO_2e$ per year (Worrall et al., 2011). This is because they are drained for agriculture or forestry – peat is removed to be used as fuel or fertiliser; they are burned, over-grazed, eroded or wasted (ibid.; Bain et al., 2011). Less than 20% of UK peatlands are currently undamaged (Littlewood et al., 2010).

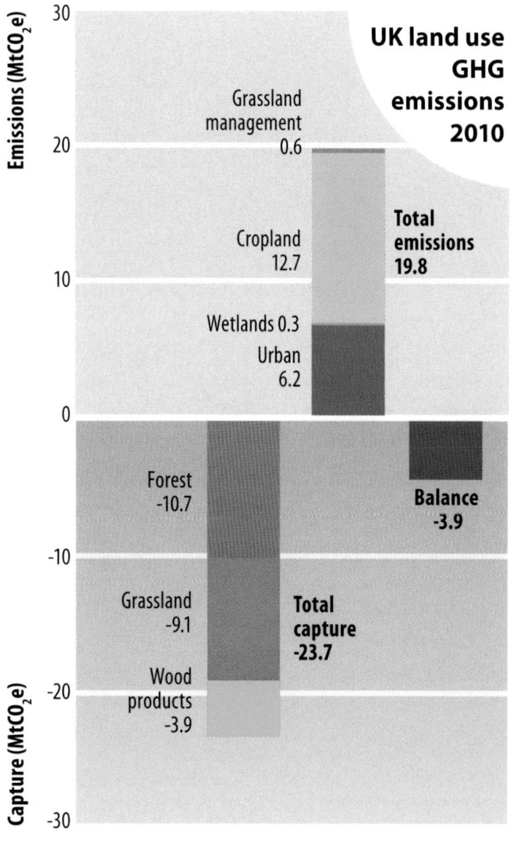

Figure 3.31: UK GHG emissions (due to land management and land use change) and carbon capture in the UK in 2010 (DECC, 2013). Emissions due to the conversion of land to cropland and urban areas are discussed in *3.6.1 Agriculture, food and diets* and *3.5 Non-energy emissions*.

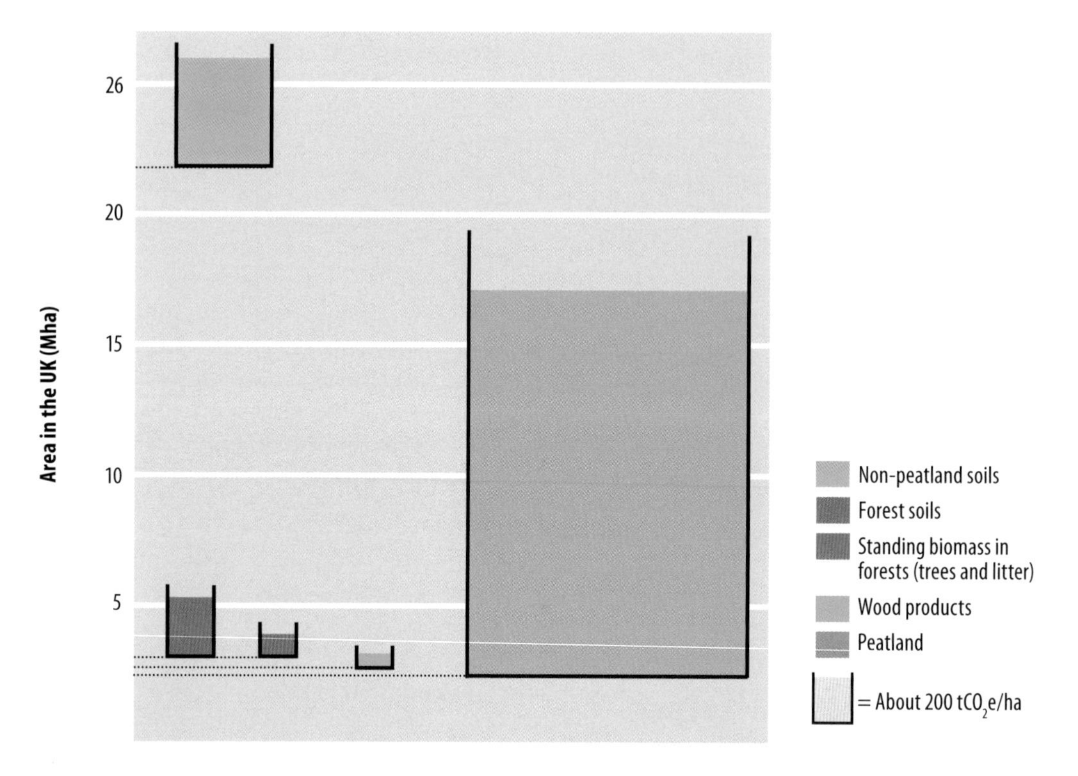

Figure 3.32: Estimated relative sizes of various UK carbon stores, measured per hectare, and the current area of each in the UK. The area used for UK wood products is that of current harvested forest area in the UK.

Balancing GHG emissions today

It is not possible to balance all our GHG emissions today simply by capturing the same amount of carbon as we emit. There are limits to how much carbon can be captured every year. The Intergovernmental Panel on Climate Change (IPCC) notes that, "only a fraction of the reduction [in emissions] can be achieved through sinks [that capture carbon]" (IPCC, 2007).

We would need a forest at least double the size of the UK to balance all our current GHG emissions (Broadmeadow and Matthews, 2003). We have to reduce emissions alongside capturing carbon.

Furthermore, even this forest would not capture carbon forever – the store would eventually 'fill up'. Most methods of capturing carbon don't last forever and are, therefore, simply 'buying us time' to replace activities which emit GHGs with alternatives that don't (Smith, 2008).

What's the solution?

The carbon cycle naturally contains a number of carbon 'flows' and 'stores'. Flows occur when carbon is added or removed from a store; stores can be built up or emptied in this way – the aim of carbon capture methods is to build up stores.

Some stores can, however, get 'full up'. How much carbon these can ultimately hold varies between different stores. Figure 3.32 shows a comparison of different UK carbon stores, though they are not necessarily 'full' yet. How fast carbon can flow into and out of stores also varies.

As a general rule, it is much easier (and quicker!) to empty a store than to build it up. The fossil fuels we are currently burning are stores of carbon that have taken hundreds of millions of years to build up (Smith, 2008). We are currently 'emptying' them over just a few hundred years (Andres et al., 1999).

Figure 3.33 shows various stores and flows of carbon. The aim here is to build up stores of carbon,

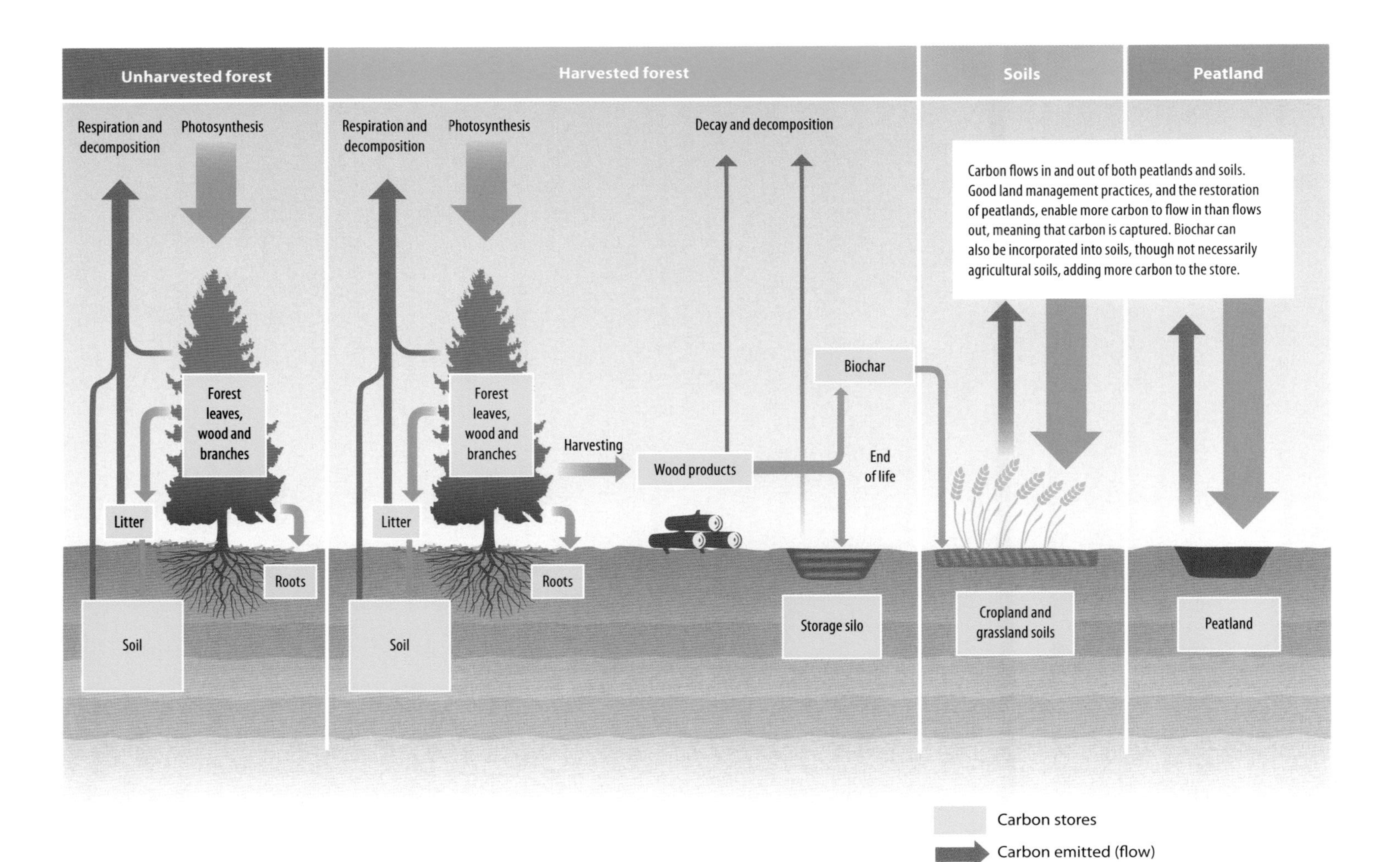

Figure 3.33: Carbon stores and flows. Adapted from Dewar and Cannell (1991) and Morison et al. (2012).

or promote continuous flows to long-term carbon stores. This can be done in a number of ways that offer:

- **One-off opportunities** where the store may become full.
- **Long-term opportunities** where carbon can continuously flow into a store.

Interestingly, the long-term opportunities usually include the very first stage (of the incredibly long process) involved in creating fossil fuels – oil, coal and gas all originated from plant and animal material. Most carbon capture processes tend to be slow, or need to be implemented on large scales to have substantial year-on-year carbon capture potential.

One-off opportunities

Different techniques of capturing carbon can last for different amounts of time before a 'store' becomes full – from a few decades to a hundred or so years. These are:

Planting forests: Forests capture carbon by taking in CO_2 during photosynthesis and releasing carbon naturally through respiration (Broadmeadow and Matthews, 2003). As a new forest grows, more carbon is captured than is released every year – it is stored in the leaves, roots, wood and branches. Some of this is released as parts of the tree die, but some remains in the tree trunk, roots and branches ('standing biomass'), or is transferred to the soils – another carbon store (ibid.). Eventually, when the forest is mature, the carbon captured every year is roughly equal to the carbon emitted over the same period, and thus the forest carbon store is full.

Planting new forest increases carbon stores over a period of 50-150 years depending on tree type. An unharvested forest can hold (or store) significant amounts of carbon once mature – up to about 1,400 tCO_2e per hectare in standing biomass (Morison et al., 2012). Planting forests can also help increase biodiversity, improve flood management (Atkinson and Townsend, 2011), and give us more natural spaces to enjoy.

Harvesting and using wood: When wood in a forest is harvested (clear-felled, or thinned) it makes space for new trees to grow and more carbon to be captured (Broadmeadow and Matthews, 2003). For this to happen the forest must, however, be sustainably managed (replanted when felled, protected from damage to soil or water, secured against illegal felling, and subject to good forest management). Harvesting wood from non-sustainable forests simply empties their carbon stores, which emits GHGs into the atmosphere (Read et al., 2009).

Carbon is stored in both the standing biomass (tree trunks, roots and branches) and in the harvested wood. Once the forest is established, up to an average of about 460 tCO_2e per hectare is held (or stored) in standing biomass, and 150 tCO_2e per harvested hectare in wood products (Morison et al., 2012). How much is in each category depends on how quickly the trees grow and how regularly they are harvested – the balance between the two stores can be different at different points in time. It generally takes between 50-100 years to accumulate the carbon – less time than unharvested forests, which are left longer to mature.

The ultimate size of the carbon store in harvested wood products depends on how many things can be made out of these materials and how long they last. Currently about 80 $MtCO_2e$ is estimated to be stored in wood products in buildings and infrastructure in the UK (Read et al., 2009). However, as a net importer of timber, the UK demand for wood products is much larger than our present home-grown supply (Broadmeadow and Matthews, 2003). This affects how much of these wood products count as 'capturing carbon' in UK GHG emissions accounts (see lower box on page 103).

UK demand for wood products, regardless of the source, could be much higher than it is today. We can use much more wood, and other plant-based materials such as hemp and straw, in construction and retrofitting. These kinds of buildings are currently unconventional, but timber-frame buildings are becoming more widespread – 22% of new builds in the UK are currently timber-framed compared to only 7% in 1997 (Read et al., 2009). If we used plant-based materials as much as possible in buildings and infrastructure, employing what are currently considered unconventional building methods, it is estimated that a massive 22 $MtCO_2e$ could be added (the net total of 'materials in' minus 'materials out') to buildings and infrastructure in the UK every year (Sadler and Robson, *undated*). This would mean big changes to the construction industry, and to the types of buildings we are used to seeing (see box to the right).

Replacing conventional building materials (steel, for example) with plant-based materials would also decrease GHG emissions related to energy use in production or industrial processes (see *3.2.1 Buildings and industry*).

Changing land use and agricultural management of soils: Soils currently store huge amounts of carbon in the UK – about 18,000 $MtCO_2e$; over a thousand times more than forests, but spread over a much wider area – see figure 3.32 (Ostle et al., 2009). Good land management can increase the amount of carbon in soils (West and Six, 2007), but most 'fill up', too. How much carbon soil can hold depends on what it is used for, and what the climate is like (Ostle et al., 2009). This means that climate change poses real risks for soil carbon stores in the UK (ibid.).

It also means that changing land use can involve a trade-off. For example, if we were to plant forest or energy crops on some semi-natural grasslands, carbon would be captured by the trees as they grew, but some would be lost from soil stores (Bell et al., 2011). Generally, semi-natural grassland soils hold more carbon than forest soils, while forest soils can hold more carbon than intensively grazed or fertilised grassland and cropland soils (Ostle et al., 2009). Therefore, converting semi-natural grassland or forest to cropland, or intensively grazed or fertilised grassland, should be avoided; and conversion to less intensively managed or semi-natural landscapes should be encouraged.

How can we use more plant-based materials in buildings and infrastructure?

Although fairly conventional building techniques (timber-frames and cladding, for example) across the building stock of the UK would increase our store of carbon substantially, plant-based materials can be used a lot more in construction in other ways. Hemp and lime offer good alternatives to conventional plaster and render, and can also be used for floors or insulation. Straw can be used in construction, too. If we use more of these materials, however, they must be sustainably sourced and made into biochar or put in silo stores at the end of their lives so that carbon is not re-released into the atmosphere.

What's important about importing wood?

About 85% of the timber we use in the UK is currently imported (Broadmeadow and Matthews, 2003). Using 'production' GHG emissions accounting, we do not take responsibility for carbon emissions *produced* on our behalf elsewhere in the world (for the 'stuff' we import), and so we cannot claim the benefit of carbon *captured* on our behalf either – trees grown to produce the timber and wood products we import. If we were to look at our 'consumption' emissions, however (see *3.10.3 Carbon omissions*), this imported timber would count towards our capacity to capture carbon. We estimate that current imported timber would constitute roughly an additional 22 $MtCO_2e$ captured per year, as long as the wood products came from sustainably managed sources.

Similarly, when looking at the 'end-of-life' of wood products, only those originally from timber grown in the UK 'count'. Again, if we were to look at consumption emissions, we could count captured carbon in imported products that were eventually made into biochar or put in 'silo storage', too – potentially another 20 or so $MtCO_2e$ of carbon captured every year.

Whether or not a soil is actively capturing carbon or releasing it depends on how it is managed (ibid.) and what the state of the soil is to begin with (Groenigen et al., 2011). We can do many things to encourage soil carbon capture in agricultural soils (and those that produce energy crops), though the soils might only continue capturing carbon for a few decades:

- On some grassland we can improve plant diversity (growing more different species) and better manage fertiliser use and grazing, resulting in up to 0.9 tCO_2e per hectare captured every year (Bellarby et al., 2013; Dawson and Smith, 2007).
- On some cropland we can reduce or stop tilling, apply manure, slurry, sewerage sludge, straw or compost, and better manage fertiliser and water use, resulting in up to 3.12 tCO_2e per hectare captured every year (Smith et al., 2000; Smith et al., 2008).

It is important these practices are maintained, or else soils start releasing carbon again. We have to look after soils over the long-term in order to keep carbon locked up (McCarl and Sands, 2007), even when the stores are 'full'. Good management practices can also mean more productive farms, however (Moran et al., 2011).

Long-term opportunities

These techniques are not time-limited and could potentially continue for thousands of years, but the rate at which carbon can be captured (how much per year) is still limited.

Peatland: Peatland doesn't behave like other types of soil. It does not get 'full up' – under the right conditions (mainly maintaining waterlogged conditions) it continues to grow. In the UK, some peatlands have been capturing carbon for over 10,000 years (Bain et al., 2011).

There is about 2.3 Mha of peatland in the UK, which currently stores a huge 19,300 $MtCO_2e$ (Ostle et al., 2009) – more than all the carbon stored in all

Biochar – we have to be careful

Although we mention that biochar may have good effects on the ability of soils to hold nutrients and water, we don't apply it to agricultural soil in our scenario. There are still some concerns over whether or not it is safe because some materials that might go into biochar production (for example, construction waste) may contain toxic chemicals. This could stay in the biochar and go into soils used for growing food. In our scenario, biochar is only added to non-agricultural soils.

Carbon in soils – we don't have all the answers

It is possible that more carbon could be captured in the long-term in soils due to either better management techniques or in forest and semi-natural grassland soils. However, how carbon builds up in soils is not well understood. Furthermore, it depends on a number of factors that aren't included in our scenario – the exact type of soil, and the current levels of carbon in it. And so, to be on the safe side we don't include any long-term soil carbon effects, but there may be substantial opportunity here to capture even more carbon.

other soils in the UK together. This store can only be maintained, and increased, if we look after peatland. Healthy peatland can capture roughly 1.1-2.6 tO_2e per hectare every year (Bain et al., 2011), but damaged peatland (particularly peatlands that have been drained) must be restored in order for it to capture this amount of carbon. This might involve changing livestock grazing, burning practices, or blocking ditches to 're-wet' drained peatland. Small-scale restoration can have an effect in as little as five years, whereas much larger interventions can take between 20-50 years to take full effect (ibid.).

If we restored all of our peatland, we could avoid emitting 3.7 $MtCO_2e$ every year, and instead capture roughly 4.2 $MtCO_2e$ every year (ibid.). This would also improve biodiversity and clean water (ibid.).

Biochar: Biochar can be made from plant-based materials, such as wood, grasses, and biodegradable wastes. It is made by 'pyrolysis' – heating at high temperatures without air (Sohi, 2012). Biochar contains a proportion of 'stable carbon' that does not degrade for a long time – from a few hundred to a few thousand years (Hammond et al., 2011).

If buried in soils, biochar can improve water and nutrient retention, helping land to be more drought-tolerant (Parliamentary Office of Science and Technology, 2010), and reducing the need for fertiliser (Sohi et al., 2010; Shakley and Sohi, *undated*), though there is still some uncertainty about this (see upper box to the left). Biochar could replace the peat currently used for soil improvement and so reduce the demand for peat from agriculture (Verheijen, 2009).

Biochar production also creates liquid and gas by-products, which can be used to produce energy. Overall, more energy is produced than is required to make biochar (Shakley et al., 2011).

How much biochar can be made depends on the materials available to create it, and how much can safely be added to soils (ibid.). Specially designed biochar storage units could also be made. Biochar made naturally by forest fires has been found in concentrations of up to 180 tCO_2e per hectare (ibid.).

Converting landfill to silo storage: Presuming every effort is made to capture GHGs from landfill, and that landfill sites are converted to 'storage silos' (see 3.5.2 *Waste* for details), a proportion of all wood products in landfill remains for thousands of years (Zeng, 2008; IPCC, 2000). This proportion, therefore, represents captured carbon (Augenstein, 2001).

One estimate states that UK-grown wood products that are currently landfilled capture roughly 3.6 $MtCO_2e$ of carbon every year (Fawcett, 2002). This doesn't include any imported timber that might also end up in landfill (see lower box on page 103). If we grew more of our own timber, and used more in construction, then it is likely that more would end up in storage silos – even if it were reused or recycled first, again meaning more captured carbon.

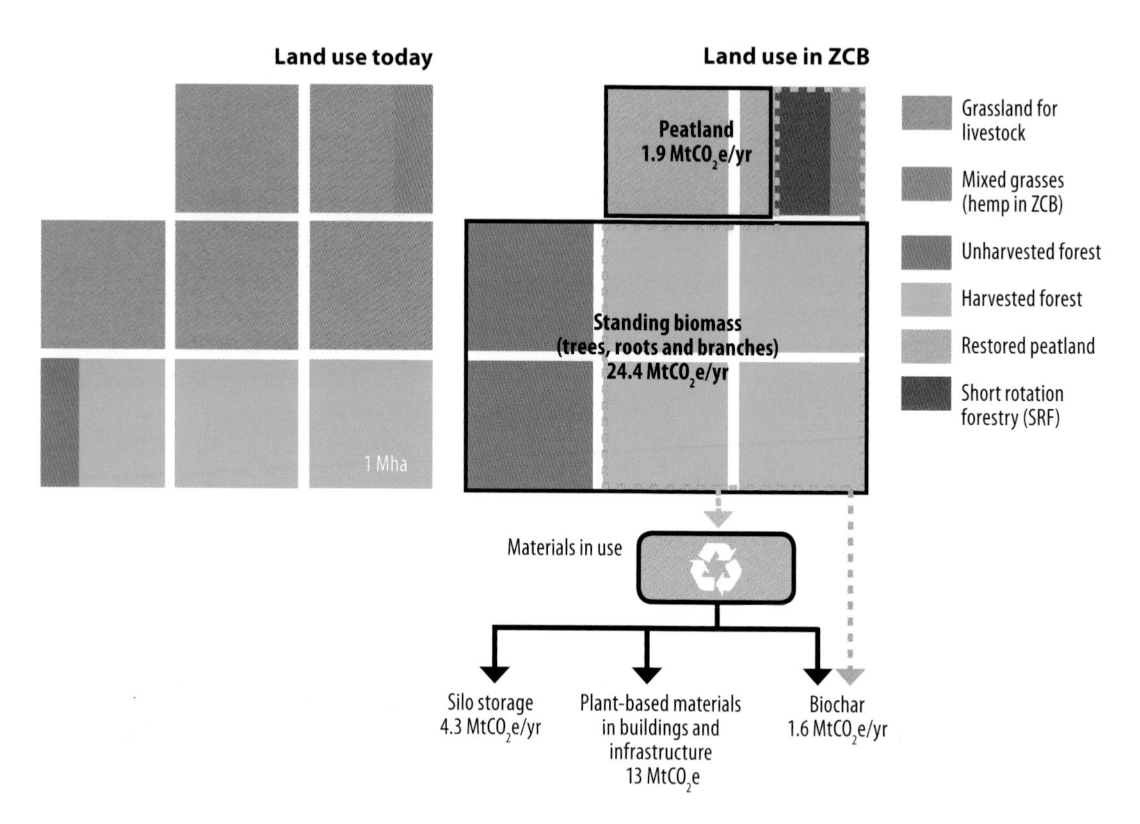

Figure 3.34: Area of land used for capturing carbon in our scenario, the methods, and how much carbon is captured as a result.

Our scenario

The ways in which we capture carbon in our scenario last for about 100 years – long enough, we think, to be able to develop other solutions for the remaining GHG emissions in the UK. The captured carbon represents an average over this time, though it may change from year to year.

In our scenario we keep land use the same as far as is possible – particularly not converting further land to cropland and losing less land to urban areas, as this is currently where the majority of GHG emissions in land use change occur (see *3.5 Non-energy emissions* and *3.6.1 Agriculture, food and diets*).

Although we don't count any long-term carbon captured by soils (see lower box on page 105), we do look at what happens in the short-term – carbon captured or released by planting woodlands and energy crops (see *3.6.2 Growing energy and fuel*), and by improved management of agricultural land. It is important that we don't lose carbon here, even in the short-term. Overall, about 240 $MtCO_2e$ is captured in soils due to land use changes (carbon is lost in some areas, but more is captured in others) and better management practices (about 12 $MtCO_2e$ per year for 20 years). Whilst this is no trivial amount, it will not continue in the long-term – after (approximately) 20 years, the soils become pretty 'full', and only very small amounts of carbon, if any, will be captured. These measures can, however, help a small amount in our transition to a zero carbon Britain (see *3.8.1 ZCB and the UK's carbon budget*). They are also likely to have benefits for the quality and productivity of soil.

To capture the carbon equivalent to our remaining impact on climate change (about 45 $MtCO_2e$ per year) we:

- Keep all of the forests that we have currently (about 2.6 Mha of harvested forestry and 0.3 Mha of unharvested forest), and manage them sustainably – replanting trees, and looking after ancient woodlands.
- Use some intensively grazed and semi-natural grasslands 'freed up' by reduced levels of livestock to:
  - Plant an additional 3 Mha of forest (doubling the forest area in the UK). 1.6 Mha of this is unharvested – simply there for us to enjoy and to enhance biodiversity, and 1.4 Mha is harvested for wood. This makes a total of 24% of the UK land forest, closer to the EU average of 37% (Atkinson and Townsend, 2011).
  - Plant about 0.5 Mha of Short Rotation Forestry (SRF) to be harvested for wood for materials (additional SRF is planted to produce biomass for heating demand – see 3.6.2 *Growing energy and fuel*).
- Plant hemp on 0.2 Mha of temporary grassland for use in buildings and infrastructure.
- Do not further degrade any of the UK's peatland, but instead restore about 50% (about 1.15 Mha).

A small fraction of wood (5%) goes to making biochar. Most of the wood products however, go into construction, and hemp is also used in buildings and infrastructure. Because of the additional use of UK-grown timber, there is additional construction and demolition wood waste in the UK's system. About a third of all this construction and demolition waste is made into biochar, and the remaining two-thirds goes into silo storage.

How all of these measures fit together is shown in figure 3.34. Carbon is captured as follows:

- 24.4 $MtCO_2e$ on average per year in standing biomass in newly planted forests (harvested and unharvested).
- About 13 $MtCO_2e$ on average per year in plant-based products harvested and used in buildings and infrastructure.
- 4.3 $MtCO_2e$ per year in silo storage.
- A net capture of 1.9 $MtCO_2e$ per year in peatlands (although some peatland is restored, the rest will still emit an amount of GHGs).
- 1.6 $MtCO_2e$ per year in biochar, added to about 0.8 Mha of non-agricultural soils.

This adds up to 45.2 $MtCO_2e$ – exactly what is required to balance our remaining impact on climate change in our scenario.

Our scenario is now net zero carbon.

3.7 Measuring up 2030

Having described the UK in our Zero Carbon Britain scenario, we can now see that many things have changed by 2030. One important implication of the changes is that we have now completely integrated our three principal metrics:

- Greenhouse gas (GHG) emissions.
- Energy supply and demand.
- Land use.

Figures 3.35-3.37 summarise our scenario for the UK.

Most importantly, our GHG emissions have decreased from 652 $MtCO_2e$ to just 38 $MtCO_2e$ – a reduction of 94%. Our remaining effect on climate change is equivalent to 45 $MtCO_2e$ in total.

Relatively small reductions in GHG emissions have been made in non-energy emissions from households, business and industry and from waste management, largely through changes to industrial processes, diversion of waste from landfill and the conversion of landfill sites to storage silos. These emissions together are reduced by just over 60%.

The largest contribution to the reduction in GHG emissions is due to changes in our energy system – how much energy we use (demand) has been reduced by about 60% from 1,750 TWh today to 665 TWh through a number of energy saving measures, and also through changing the way and the amount we travel and move goods. We produce 1,160 TWh of energy to supply our needs, covering losses in the system, requirements for synthetic gas and liquid fuels and back up to balance supply and demand. This is produced completely using renewable energy and carbon neutral energy sources, meaning that GHG emissions from energy use are zero.

This system also has implications for land – in total, about 17% of our land is used to produce energy in some way, either fuel for transport and industry or as back up for our electricity system. More of our landscape is used to grow Short Rotation Forest, Coppice and various grasses for energy production – a significant change to the grazed fields we are used to.

GHG emissions from agriculture have decreased substantially – by roughly 73%. This is largely due to changes in our diet, including significant decreases in the amount of meat and dairy we eat, plus changes in management practices and the elimination of the need to use ever-more land for agricultural purposes. In total, we now only use a third of our land to feed ourselves (compared to 70% today), despite importing less food from abroad (about 17% of the food we eat is imported, compared to about 42% today). Over half of our agricultural land is still dedicated to livestock (sheep and cows) in some way – either grazing grassland or growing feed.

Another significant change to our landscape is a doubling of the area of forest. A larger proportion of this – 30% – is unharvested, meaning there is more space for biodiversity. A larger proportion of land in the UK (almost 15% compared to only 8% in present day UK) in not used productively, increasing the space for wild, conserved or protected areas, including restored peatlands – all of which are very important habitats for biodiversity, not just for carbon management.

All together, these changes to the way we use land, the increased area of forest, the restoration of 50% of our peatlands, and the use of more plant-based products made mainly from harvest wood, allow us to capture about 45 $MtCO_2e$ every year.

This balances out the emissions left in our scenario, meaning that we capture the amount of GHGs equivalent to our remaining impact every year – we are net zero carbon.

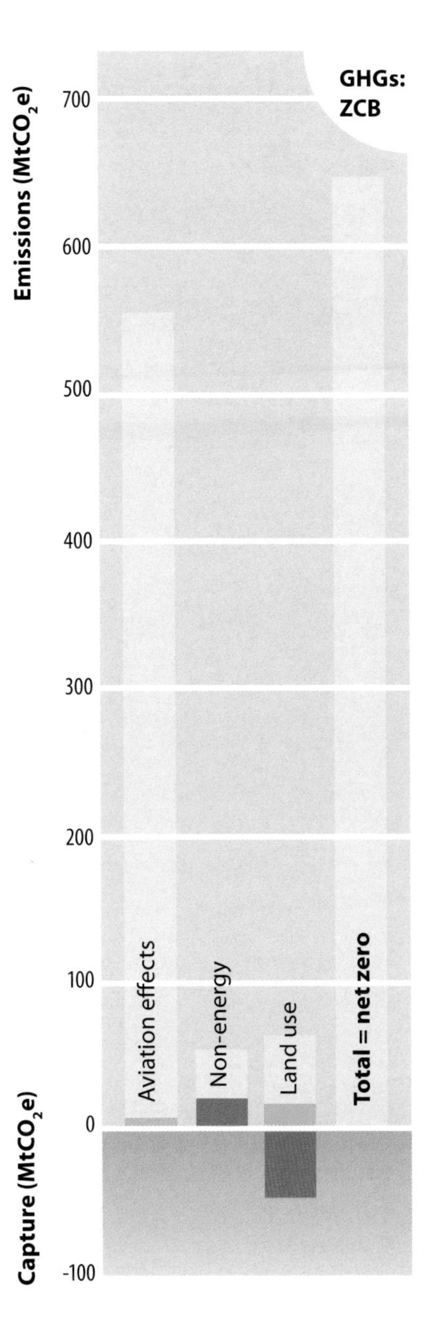

Figure 3.35: Carbon captured and greenhouse gas emissions for the UK in our scenario relative to 2010, including international aviation and shipping and the enhanced effect of emissions from aviation. Total emissions sum to net zero.

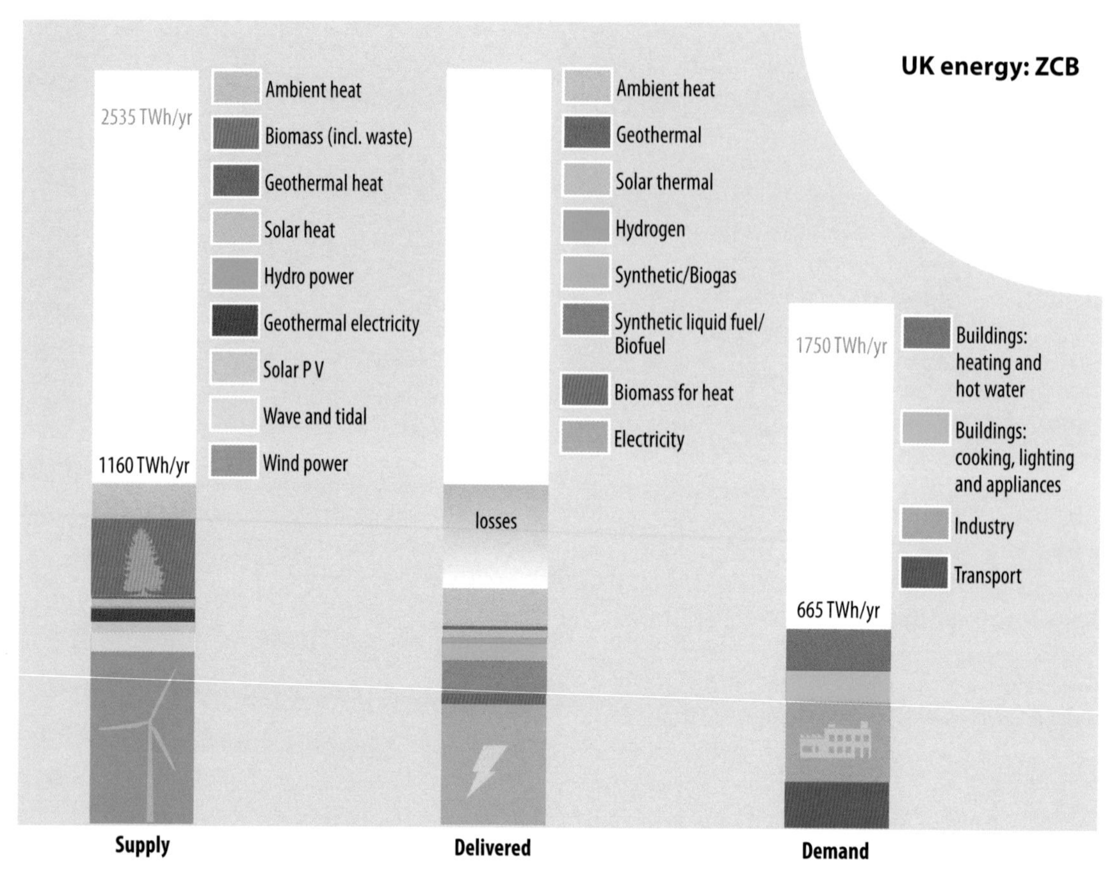

Figure 3.36: Primary energy supply, delivered fuel mix, and final energy demand for the UK in our scenario, relative to 2010.

Figure 3.37: Approximate land use in our scenario (not including water courses and coastal areas). 'Mixed grasses' includes hemp, Miscanthus and other energy grass crops.

3.8 How we get there

Our scenario describes what it could be like if we in the UK rose to the challenges of the 21st century. We've taken this approach not because the ways to get there aren't important, but because we need to know where we're going in order to face the magnitude of the challenge ahead.

We don't model the transition in detail, but it is interesting to discuss some of its implications – what challenges this might pose for policy, and what opportunities for our economy there might be along the way.

3.8.1 ZCB and the UK's carbon budget

In 2.3.1 *Our carbon budget*, we estimated the UK share of the global carbon budget as:

- 8,400 MtCO2e (corresponding to an 80% chance of avoiding a 2°C global average temperature rise).
- 9,600 MtCO2e (75% chance).
- 11,200 MtCO2e (67% chance).
- 14,000 MtCO2e (50% chance).

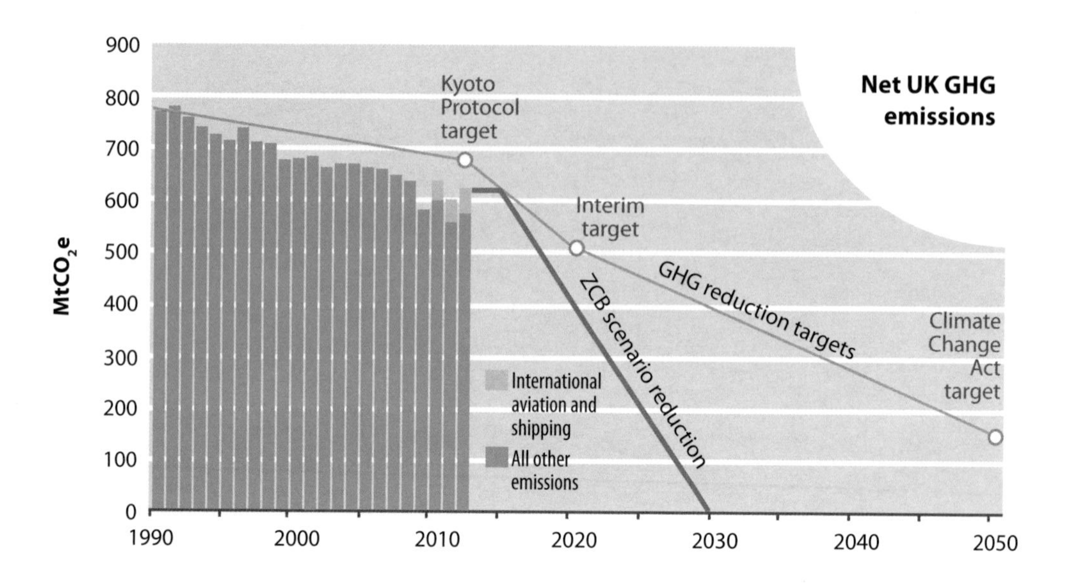

Figure 3.38: Transition used to estimate the total carbon 'spent' in transition to our ZCB scenario, modelled for our scenario relative to current UK policy. Adapted from Beales (2013).

Using data on UK GHG emissions from 2010 to 2012 (DECC, 2013b and DECC, 2013a) we then assume that up until the beginning of 2015 we continue to emit roughly the same amount of GHGs as in 2012. Beyond 2015, our emissions decrease linearly until they reach zero in 2030. Our emissions remain at net zero after 2030. Figure 3.38 shows how this transition compares to current UK policy emissions reductions targets.

Adding up our emissions year-on-year, and then deducting how much carbon we have captured in short-term measures, such as improved land management techniques (roughly 240 $MtCO_2e$ – see *3.6.3 Capturing carbon*), tells us how much carbon we 'spend' over the period 2010 to 2050.

What about emissions resulting from changes to UK infrastructure?

What we may not have accurately represented in our simple linear decarbonisation is the additional carbon we might 'spend' (or emit) in building the infrastructure in the scenario. Materials for those offshore wind farms and insulation for our houses all have to be manufactured and transported, and until they are in place the energy we use to do this will cause GHG emissions.

This might make the shape of the transition quite different. For example, there could even be an increase in emissions at the beginning (as we get busy building), followed by a sharp decrease in emissions once the majority of our energy supply becomes 'zero carbon'. Without modelling this transition fully, we can't really say whether we would 'spend' (or emit) more GHGs in total over the whole period, or less. This is an important avenue for further research.

What about our historical responsibility?

How much 'historical resonsibility' we take for GHGs we have emitted in the past is an important and difficult moral issue that requires substancial attention from international policymakers. Since most GHG emissions persist for hundreds of years, a substantial amount of what is now in the atmosphere is 'ours'. In some sense, we may have already exhausted our 'moral budget' – having emitted far more than our 'fair share' over the years since the industrial revolution. For example, a large proportion of the 400 $GtCO_2e$ of GHGs emitted globally between 2000 and 2009 (FoE, 2010) were from industrialised nations like the UK – far more than our 'fair' per capita share. The division of a global carbon budget on a per capita basis from an earlier date therefore means that we take responsibility for the fact that we have emitted more than our 'fair share' of global emissions *in the past*, as well as continuing to do so today.

Between 2000 and 2009, UK GHG emissions came to about 6,800 $MtCO_2e$ in total (almost double our 'fair' per capita share of the GHGs emitted globally), meaning we have already 'spent' a larger proportion of what is available to us through to 2050. This means, under this frame of historical responsibility, between 2010 and 2050 the UK's remaining budget is:

- 5,100 $MtCO_2e$ (corresponding to an 80% chance of avoiding a 2°C global average temperature rise).
- 6,300 $MtCO_2e$ (75% chance).
- 7,900 $MtCO_2e$ (67% chance).
- 10,700 $MtCO_2e$ (50% chance).

Comparing this again to the 7,450 $MtCO_2e$ 'spent' between 2010 and 2050 in the transition to our scenario, we now find that we only come in 'under budget' for the UK share of a global carbon budget corresponding to a 67% chance of avoiding a 2°C temperature rise – we have 'overspent' budgets with better chances. This is still far better than current UK policy, but is a one in three chance of what is now defined as 'extremely dangerous' climate change too high?

There is also the question, what *is* 'fair' – how far back should our emissions be counted?

The longer the frame of historical responsibility we take (the further back we go), the harder it is for the UK – and other long industrialised nations – to keep to a budget that gives any reasonable chance of avoiding a 2°C temperature rise. However, there are some options open to us in these cases:

- Faster decarbonisation. This means we tighten our purses and 'spend' less carbon. For example, for an 80% chance in the above example, the UK would have to fully decarbonise before 2020.
- More carbon capture. This means we rein in our 'overspending' by 'earning' more. It would be beneficial to maximise techniques that capture carbon – those that work in both the short-term and long-term are beneficial here (*3.6.3 Capturing carbon*). Other geoengineering options to remove CO_2 from the atmosphere may also be considered (*3.1 About our scenario*) should these methods be exhausted.
- International credits. This means we pay others to cover our 'overspend'. Paying for our remaining emissions and funding the transition to zero carbon economies in less developed nations has been recognised as an important aspect of global decarbonisation (Chichilnisky, 1994).

The latter two options do not provide alternatives to rapid decarbonisation, but are complementary – they 'settle up' historic contributions to the problem and 'buy us more time' for the process of decarbonisation. There are limits to how fast we can decarbonise, but also to how much carbon can be stored, and to how many credits it would be possible to purchase in an equitable and effective scheme (UNEP, 2012). We cannot rely on any of the above options individually. What is clear is that as a long industrialised nation, we have a responsibility to cut our emissions to net zero as quickly as possible.

In all cases, rapid decarbonisation is necessary, but it may not always be sufficient. Historical responsibility is an important question that can only be addressed at an international level, and will play an important part in future climate negotiations.

This comes to about 7,450 $MtCO_2e$ including emissions from international aviation and shipping (currently not counted under the UNFCCC Kyoto Protocol).

This very simple estimation implies that we come in 'under budget' for the UK share of a global carbon budget corresponding to an 80% chance of avoiding a 2°C temperature rise, when current UK policy fails to meet the criteria even for a 50% chance (*2.3.1 Our carbon budget*).

It may even seem as though our scenario has some budget left to 'spend', but this is misleading. Global cumulative carbon budgets do not represent 'hard limits', but a sliding scale of risk. Essentially, the sooner we decarbonise, the smaller our contribution to the problem and the better our chances of avoiding what is now defined as 'extremely dangerous' climate change (Anderson and Bows, 2010). Acknowledging our historical responsibility as a long-industrialised nation only further emphasises the necessity to decarbonise rapidly: to help international negotiations and catalyse global action on climate change.

3.8.2 Zero carbon policy

Strong policies to enable rapid decarbonisation – those that reduce GHG emissions quickly and equitably – are urgently required. Such transitions are still at the very boundaries of what is politically thinkable and are as much a challenge for our policy community as for our scientists and engineers. It requires a new degree of urgency coupled with joined-up thinking on local, national and global scales.

Current policy frameworks and mechanisms

A policy framework is an organisational structure of multiple policy mechanisms working together to achieve a set of required goals. For example, the Kyoto Protocol under the United Nations Framework Convention on Climate Change (UNFCCC) is an international agreement where the policy framework comprises three mechanisms that in combination help achieve its emissions reductions targets.

Here we explore and compare a range of leading potential policy mechanisms.

'Cap' schemes are the most widely adopted policy mechanisms designed to reduce GHG emissions. A cap (or total limit) on emissions is set and tightened over time so that the GHG reductions it achieves are in line with international agreements, nationally adopted targets or organisationally determined goals. Cap schemes also involve putting a price on carbon so that the damage caused by GHG emissions is internalised – or taken into account – by economic markets. There are two types of cap schemes:

- 'Hard cap' schemes do not permit emissions to exceed the cap, meaning that the 'carbon price', and therefore potentially the price we pay for goods and services, changes over time.
- 'Soft cap' schemes do permit emissions to exceed the cap under certain circumstances, but put a high 'carbon price' on GHG emissions to discourage this.

Cap schemes can be implemented either 'upstream' or 'downstream'. Upstream systems target suppliers of fossil fuels and energy services directly; downstream systems seek to change individual behaviours, such as home energy use, driving and flying. In essence, they both treat GHG emissions as a tradable commodity ('carbon trading'). Companies or individuals who emit less than their share (as defined in various ways) can sell their surplus to those who have emitted more than their share.

Upstream systems are currently more common, but downstream systems are also being explored by policy makers. Table 3.3 outlines the advantages and disadvantages of various policy mechanisms that could help reduce the UK's GHG emissions.

Which policy framework is best for the UK?

It is unlikely that any one single policy mechanism can deliver the radical emissions reductions we now require. We will need a policy framework combining effective mechanisms designed to work well with a range of sectors, including energy production, industry, housing, business, transport, land use and agriculture. Working national and local policies together in this way has been shown to be effective in reducing emissions (UNEP, 2012). Closing the gap between what is physically necessary to address climate change and what current UK emissions reduction targets are projected to achieve (see *2.3.1 Our carbon budget*) will require high-level all-party political commitment, cross-sectorial collaboration and public engagement at every level.

The next section, *3.8.3 Economic transition*, describes how some of these policy mechanisms can be used to decrease emissions on a national level, and what effects they may have for the UK economy.

Some of the mechanisms that operate on an international level are subject to 'carbon leakage' – moving production abroad to areas where carbon trading has not yet been implemented, or where carbon taxes are lower. Some mechanisms can even provide a disincentive to decarbonise, especially in the short-term, delaying decisions and leading to infrastructure 'lock-in' that commits to higher energy use or emissions over the following decades. If policy mechanisms are to be effective on a global level, they must be designed to avoid or manage these issues (ibid.). Potential solutions could be delivered through trade agreements like those organised by the World Trade Organisation (WTO), or by using border taxes to level out costs (Carbon Trust, 2010). This issue is associated with the need for us to take account of emissions arising from the production of goods that we import, and not just those from goods produced at home, which is currently the case in international agreements under the UNFCCC (*3.10.3 Carbon omissions*). Taking full account of emissions arising from internationally traded goods is a vital area for future policy research.

	Mechanism	Description	Advantages
UPSTREAM	Emissions Trading Scheme (ETS)	The government sets a soft cap on carbon emissions. Allowances or permits are distributed among industry and businesses. They must have sufficient permits to cover the emissions they produce, either through the initial allocation, or auction, or by trading with others.	• The emitter can emit only limited GHGs, which reduces over time. • It provides incentives for industry and businesses to develop low carbon technologies to keep their emissions within defined limits.
UPSTREAM	Cap and share	A hard cap is placed on GHGs emitted by fossil fuel suppliers. Emissions permits are distributed equally among adult citizens. Each citizen, or group of citizens, can choose to sell the permits to fossil fuel suppliers. The money raised by the sale of permits can be shared between citizens.	• The cap is enforced by requiring the fossil fuel supplier to pay for a fixed amount of emissions permits. • The money raised can compensate for a potential rise in energy (or fuel) prices.
DOWNSTREAM	Tradable Energy Quotas (TEQs)	A hard cap is set on emissions and an annual carbon budget is set based on the speed of emissions reduction required. The proportion of emissions associated with households is distributed equally to every adult citizen for free. The remaining permits are sold by tender to all non-household energy users. All fuels would carry carbon ratings. A consumer 'pays' in carbon permits to cover the rating of their purchase (Fleming and Chamberlin, 2011).	• There is equal entitlement to fuel use among all citizens. All other energy users are also included. • It provides large incentives to all sectors of society to reduce their carbon emissions. • All emissions from energy use can be measured simply and efficiently by assigning a rating based on the quantity of GHGs generated by their production and use. This avoids the need for complicated lifecycle emissions calculations. • It can help deal equitably with restricted energy use as well as emissions reduction.
DOWNSTREAM	Personal Carbon Allowances (PCAs)	Emissions allowances are allocated equally to adult citizens (half an allowance is proposed for children) and can be 'spent' as required on GHG emitting activities, such as paying a gas bill. Those who keep to budget will have spare quota to sell, whilst those who don't will have to buy allowances to cover their excess.	• Individuals can either maintain existing behaviours and buy allowances, or change their behaviour and reduce their emissions, potentially profiting by doing so. • There is the potential to constrain emissions in 'an economically efficient, fiscally progressive, and morally egalitarian manner' according to some (Roberts and Thumin, 2006).
TAXATION	Carbon tax	A tax is imposed on the release of GHG emissions from industry and businesses, providing an incentive to reduce GHG emissions if doing so costs less than paying the tax.	• Simple to design and implement. • Raises the cost of using fossil fuels and encourages innovation and investment in developing renewable technologies and more energy efficient processes.

Table 3.3: A comparison of different policy mechanisms.

Disadvantages	Recent developments
• The 'carbon price' (the cost of each permit) can fail to reflect the real cost of environmental damage in the long-term. • It can be cheaper to buy permits from other businesses, rather than reducing emissions – especially in the initial stages of the scheme. • Can be subject to 'carbon leakage'.	• The European Union (EU) ETS is the world's largest carbon market and is now in its third phase (2013-2020). • National or sub-national systems are being operated in Australia, Japan, New Zealand and the USA, and are planned in Canada, China, South Korea and Switzerland.
• Administrative and commercial systems, such as banks or post offices, are needed to support the operation of the scheme. • The ability of the citizen to make money out of the scheme may reduce the motivation to achieve emissions reductions. • Can be subject to 'carbon leakage'.	• 'Cap and Dividend' is a similar mechanism and has gained political popularity in the USA.
• Administrative systems are required for the registration of permits and all transactions. • Can be subject to 'carbon leakage'.	• The TEQ concept has been embraced in France, and the Resource Cap Coalition is assigned to carry the idea across Europe. • It has won support from the main political parties in the UK. A policy framework for peak oil and climate change was published in 2011 with support from all parties.
• The mechanism applies to individuals and may have limited impact on the economy as a whole. • Administrative systems are required for the registration of allowance quotas and all transactions. • There is potential for unequal effects on individuals – a recent study suggested households in rural areas, detached houses, or those that use oil and electricity for heating, retired people or single dwellers without children, are more likely to experience a deficit of PCAs (White et al., 2013).	• This mechanism is only in the research phase.
• There is potential for unequal effects on individuals – increased production costs caused by the carbon tax may be passed on to consumers, having a larger impact on low-income households. This could be addressed through subsidies. • There is no guarantee that the tax would keep emissions within the carbon budget. • Setting the 'right' carbon price that would change behaviour sufficiently to avoid emissions is difficult.	• The EU plans to phase out all subsidies for fossil and nuclear energy and introduce an EU-wide carbon tax by 2050 (EREC, 2010). • The Climate Change Levy (CCL) is the carbon tax currently in use in the UK. It only applies to energy used for lighting, heating and energy in non-domestic sectors. • The Carbon Price Floor (CPF) has recently come into force as a tax on fossil fuels used in parts of the energy sector in the UK.

3.8.3 Economic transition

We now have a chance to change everything, because everything must be changed. Reducing our debt burden is of course an important part of an economic recovery plan, but dealing with recession through cuts alone is not working. Billions are being spent in schemes designed to kick-start our stagnant economies, but most are focused on a return to 'business as usual'.

Growth in key sectors can form part of an economic recovery, but it must also now embrace a transition that keeps us within the limits of global ecosystems. By making visionary investments at ground level, we not only create employment and stimulate the economy, but we also 'future-proof the UK' to be ready for the climate and energy challenges of the 21st Century.

Britain's economic reliance on financial services and a consumer retail economy are still too high – we need to rethink the economy, based on harvesting our natural assets and valuing our ecosystems. To this end, the Zero Carbon Britain project has been exploring links to projects such as the 'Green New Deal', which has been working to develop this type of recovery programme for the UK economy.

The Green New Deal approach is a cornerstone of new economic thinking that will move society on from doing the things that got us into so much trouble in the first place. It will not only drive the kind of powering down and powering up measures outlined by the Zero Carbon Britain project, it will break the bonds of fossil fuel dependence, protect against the economic impacts of fossil fuel price rises, create employment, tackle fuel poverty, generate jobs and promote a dynamic, modern, low carbon economy. The large part of the investment in renewables is made upfront at the time the technologies are installed, and as such is more predictable and quantifiable.

So, rather than paying to import from a peaking pipeline of polluting fossil fuel imports, an 'energy lean' British economy can be driven by an indigenous renewable energy supply chain. By their very nature, these renewable reserves will not peak.

The creation of a resilient zero carbon economy is a vital step towards the important end goal of a 'steady state' economy. By learning the hard economic lessons of recent years we can stay ahead of events and refocus the ingenuity of the finance sector on the actual challenges at hand, creating a new kind of energy-lean, decarbonised economy: stable in the long-term, locally resilient, but still active in a global context; rich in quality jobs, a strong sense of purpose and reliant on indigenous, inexhaustible energy. This is an important area for further research.

The Green New Deal

Taking inspiration from Franklin D. Roosevelt's original New Deal, which drove the recovery from the 1929 economic crash, the Green New Deal outlines an integrated response to our current economic challenges through a dual approach:

- Firstly, it entails sorting out the rules by which the economy works, so the problems will not re-occur. This means re-regulating finance and taxation so finance will return to its role as servant, not master, of the economy: dealing prudently with people's savings and providing regular capital for productive and sustainable investment.
- Secondly, it outlines a transformational programme using a mix of public and private investment to restart the economy in a way that will both rapidly decarbonise it to meet the climate challenge, and also create employment by tapping into the UK's massive renewable energy asset base.

To tackle the problems facing us at the speed required we need the equivalent of an 'environmental war effort' – the Green New Deal offers a path to re-engineer the economy at a scale and speed only previously seen during wartime (Green New Deal Group, 2008). This approach will deliver huge increases in investment in energy infrastructure backed by a new legislative framework offering price signals adequate to accelerate the shift away from a fossil fuel based economy. These signals should include rising carbon taxes and a price for traded carbon that is high enough to cause a dramatic drop in carbon emissions (ibid.). Once the market becomes alert to the economic role of carbon, the most economically effective option automatically becomes that with the lowest embodied emissions, and the economy itself becomes an engine for rapid change – effectively, a race away from carbon. Economic market pressure for ever-lower carbon options then accelerates the development and implementation of new kinds of technologies.

The Green New Deal can be driven by a mix of public and private finance:

- The public funding for the Green New Deal would come in part from the increase in the Treasury's coffers from rapidly rising carbon taxes and carbon trading. Also, now that energy prices are high, and before North Sea oil is exhausted, introducing a windfall tax on oil and gas companies would be a huge funding source. Fossil fuels are an unrepeatable windfall from nature, yet the UK Government has so far failed adequately to take advantage of its income to prepare us for a low carbon future. Norway, by contrast, has used its oil surpluses to help create an investment fund for future generations that is today worth around €260 billion (£198 billion). This amounts to €75,000 (£57,000) for every man, woman and child in the country. The UK could follow Norway's lead and set up an Oil Legacy Fund, paid for primarily by a windfall tax on the current high level of profits of oil and gas companies (ibid.).
- The private funding needed for the transition could be secured by releasing the literal wall of money in pensions and other savings, which are urgently seeking secure long-term returns. Pension funds have a rising demand for relatively risk-free assets to match their liabilities. The solution lies in a new generation of Green New Deal 'climate bonds' backed by municipalities, national government and international financial institutions (ibid.).

Investment in such actions would not only inject money into the economy at ground level it would also help create a vast 'carbon army' of re-skilled workers, and offer tangible returns to repay the investor – be they the taxpayer, an individual or a pension fund – from the price of the energy saved or generated.

From the new economic foundation's Green New Deal *(Green New Deal Group, 2008).*

3.9 Benefits beyond carbon

Tommy Hansen / Public domain

Although our scenario is primarily focused on reducing GHG emissions, as we outlined in *Chapter 2 Context* there are a number of other interrelated problems facing us today. We are now seeing signs of serious environmental change, driven by global society's continued failing to live within the limits of the planet's eco-systems.

It is important not to have tunnel vision when it comes to rising to the challenge of climate change. It will require a new approach to many of our current lifestyle choices – the trick is to identify synergies within the changes that help increase our ability to adapt to climate change, our readiness for peak oil, our wellbeing, employment, economic recovery, biodiversity and many other things.

Overall, on a very basic estimation, we think our scenario actually has mostly positive implications. Many of the changes to the UK and to our lives in our scenario also:

- Help us adapt to what we now accept to be inevitable changes in our climate. There is, however, a risk that this adaptive capacity, and perhaps even our ability to mitigate effectively, would break down should the changes become more extreme.
- Address many of the detrimental trends that on a local scale cause us to surpass the 'planetary boundaries' that define a 'safe operating space for humanity'– for example, the loss of biodiversity. They also help address global issues, such as ocean acidification.

- Help create new employment opportunities in the UK in many sectors, to the tune of roughly 1.5 million jobs.
- Help foster a society in which we are happier and our wellbeing is increased, with a greater sense of collective purpose.

3.9.1 Adaptation

Section *2.1.2: Climate change* describes some of the signs that we can now see of climate change starting to bite. Moreover, the impacts that were associated with exceeding a 2°C temperature rise have been revised – 2°C now more appropriately represents the threshold between 'dangerous' and 'extremely dangerous' climate change (Anderson and Bows, 2010).

Our actions must embrace crisis management (adaptation) *alongside* crisis prevention (mitigation) – and this will become increasingly important. We cannot tackle either adaptation or mitigation in isolation if we want the practical solutions we propose to future-proof us against the challenges ahead.

"Climate change is already happening and is bound to continue because of the amount of greenhouse gases already in the atmosphere. Even the toughest mitigation efforts and targets cannot avoid further impacts of climate change in the next few decades. Adaptation to these impacts ... is therefore essential and of critical importance." (NCCARF, 2010)

Though regional (small-scale) impacts of climate change are hard to calculate, the UK Climate Impacts Programme (UKCIP) presents projections of what might happen in the UK with a changed climate and helps us see both the risks and opportunities (Jenkins et al., 2009). For instance, it projects:

- Warming for the UK over the long-term, more so in summer than in winter.
- The warmest days of the year to be generally much warmer than now.
- The very coldest days in winter to be generally less extreme.
- More rain in the winter and less in the summer (potentially leading to increases in both floods and droughts), though throughout the year the amount of rain that falls may remain about the same.
- Sea levels to rise and coastal waters to be warmer.
- Storms and 'extremes' of weather are likely to change, though where these types of weather events happen, how often, and how severe they will be is much less certain.

This probably isn't everything!

We haven't performed a full 'adaptation review' of our scenario, but in table 3.4 are some of the opportunities and risks that have become apparent during our research. There are gaps in our knowledge. For example, we say little about water resources, seas and oceans, as they are not explicitly modelled in our scenario. There are many links between different systems, many things we are uncertain about, and many changes we will not be able to predict. As such, any planning for adaptation would need to have flexibility and continuous reassessment based on good evidence at its core (DEFRA, 2010).

The degree to which these projected effects are likely to happen depends on how quickly GHG emissions are stopped worldwide. But we should not rely on regional projections too much – they can be uncertain, meaning we have to have a flexible approach to adaptation. Furthermore, since these projections were made, additional evidence would suggest that changes to 'extremes' in weather may be underestimated (Deser et al., 2012). These extremes are perhaps the most difficult conditions to adapt to.

Though the Zero Carbon Britain scenario reduces emissions to net zero and maintains the UK's share of a global carbon budget that increases our chances of avoiding 'extremely dangerous climate change', it

Opportunities	
Buildings and industry	Insulation in homes and offices can have a dual function – if designed properly. It can keep heat in or keep heat out, meaning that better insulation will work well in a warmer climate.
Transport	
Energy	Sea level rise and increasing storm surges increase the risk of serious disasters at nuclear power plants located in coastal areas (Greenpeace, 2007). Our scenario eliminates this risk by removing nuclear power from the energy mix. Having a flexible energy supply might help us adapt to changes in energy demand as the climate changes.
Land use	Planting trees in cities and towns can provide shade and a localised cooling effect, helping urban areas adapt to warmer conditions (Atkinson and Townsend, 2011). In rural and urban areas trees assist in better water management – helping to prevent droughts and floods (ibid.; Read et al., 2009). Planting new forests also increases biodiversity – the more diverse the ecosystem, the more resilient it is to change. Trees also provide good 'wildlife corridors' allowing animals to move easily when necessary to avoid any extreme effects of climate change, or to adapt to changes in local climates. Restoring peatlands increases biodiversity, making natural habitats more resilient to change. Peatlands can also help improve water resources (Bain et al., 2011).
Agriculture	Incorporating biochar into soils may help keep nutrients and water in, making crops more resilient to dry periods and increasing crop fertility (Sohi et al., 2010). With warmer temperatures and longer growing seasons, the types of crops that we can grow in the UK may change, providing opportunities to grow a wider variety.

Table 3.4: Opportunities within our scenario for adaptation to a changed climate; the risks a changed climate might pose for our scenario; and some management options which may help decrease the risks. Much of the information is taken from the Climate Change Risk Assessment (HM Government, 2012).

Risks	Management
Flooding has already caused much damage to UK infrastructure. More intense rainfall means a higher risk of flooding.	Many natural systems can help decrease risks of flooding (see Land use below).
Risks to transport include increased flooding and warmer temperatures (expansion of rail tracks in hot weather, for example).	Designing transport systems to cope with more extreme weather in mind may help reduce the risks.
Changing weather patterns may affect where best to locate renewable energy systems – extreme weather could cause problems for renewable energy generation – storm surges and floods may affect hydropower systems, for example.	Building in flexibility to energy generation technology can help reduce the effects of extremes and changing weather patterns – making hydropower generation technologies better able to cope with higher river flows, for example.
As temperatures increase, carbon can be lost from peatlands. The efforts to restore peatlands for carbon capture could, therefore, be reversed.	Tree species that are planted must be suitable for the climate that we expect over the next 100 years, since forests take so long to establish. Damaged peatlands are much less resilient than healthy ones (Bain et al., 2011). Peatland restoration offers the best chance of reducing the risk of carbon loss.
With potentially warmer and wetter conditions, diseases may travel better. This poses risks for crops and can cause harvests to fail. Heat stress and drought also pose risks for crops (both food crops and energy or fuel crops), affecting how well they grow. As temperatures increase, carbon can be lost from soils.	Crop species choice can be changed regularly to suit the climate as they are replanted regularly. 'Monocultures' (single species) should be avoided. Where possible, planting of mixed crop species should be encouraged. Planting crops which are more drought and flood resilient should also be encouraged. By improving management techniques of agricultural soils (on cropland and grassland), carbon loss can be minimised.

must still adapt to new 'norms' and new 'extremes' in climate – both the expected and *unexpected* changes.

We will have the opportunity to build in adaptive capacity as we make changes to the way we live. Equally, however, we must be aware from the outset that there are some very real risks that may prove difficult to overcome. Table 3.4 shows a basic appraisal of our scenario under a changed climate. It shows that our scenario increases our adaptive capacity in most cases, though extra consideration in the design of infrastructural changes would be needed.

3.9.2 Planetary boundaries

Pressures on the 'planetary boundaries' are driven by a combination of population growth, increased consumption and environmentally damaging production systems (Rockström and Klum, 2012). There is probably little that can be done to modify the trajectory of population growth, now slowing and projected to stabilise globally at around 9 billion in 2050 (Lutz et al., 2004).

With respect to increased consumption, it is important that economic growth is concentrated where it is needed most – in developing countries. The already wealthy regions (largely the Western world) need to plan for low growth and a transition to steady state economies (Victor, 2008).

With these changes as a background, our scenario proposes a wide range of technological shifts in production methods, as well as changes in consumption patterns. Although the planetary boundaries are measured globally, some of the issues depend on local conditions. For others, there are obvious links between local actions and global effects – for example, adhering to a nation's carbon budget plays a part in global efforts to tackle climate change. Whilst the UK can play its part in helping local and global conditions progress in the right

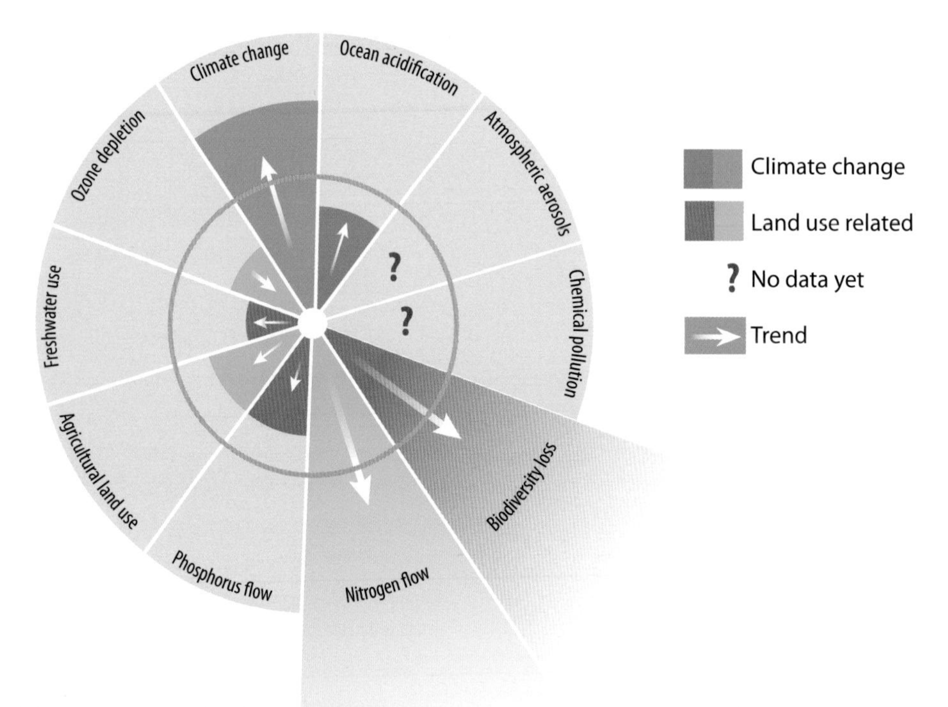

Figure 3.39: Diagram showing the planetary boundaries, divided into sections representing those which have common roots of climate change and of land use. The grey circle represents the 'safe operating space' for humanity. Adapted from Rockström et al. (2009).

Caza_No_7 / Creative Commons 2.0 Generic

directions, many of the trends ultimately depend on co-ordinated global action.

Zero Carbon Britain is focused on the **climate change boundary**. It tries to demonstrate an adequate national contribution to the planetary problem through complying with the proposed global budget for accumulated GHG emissions to 2050 (*3.8.1 ZCB and the UK's carbon budget*).

Ocean acidification is directly connected with climate change (see figure 3.39), as oceans acidify through the uptake of CO_2 from the atmosphere – the more CO_2 there is in the atmosphere, the more the oceans acidify. There are some geoengineering proposals for dealing with climate change that would leave the acidification problem unchanged – for example, shielding the Earth from the sun's heat to keep temperatures down (Williamson and Turley, 2012). Reducing GHG emissions (and thus levels of CO_2 in the atmosphere), as in the Zero Carbon Britain scenario, targets both climate change and ocean acidification simultaneously.

The pressures on the other planetary boundaries (not including Ozone depletion, which is already improving (UNEP, 2012), plus those not quantified yet) are broadly proportional to how much land we use (see figure 3.39), and how intensively we use that land. There can be little doubt that the largest driver of unsustainable trends has been increasing consumption of grazing livestock products (largely beef, lamb and dairy), which require much more land than crops (Nelleman, 2009; Pelletier and Tyedmers, 2010; Foley et al., 2011; Greenpeace International, 2012).

Therefore, the simplest way to help change these trends is to reduce livestock production and consumption (Elferink et al., 2008). This is largely what our scenario does, particularly by reducing grazing livestock significantly and banning the import of livestock products and livestock feed.

Looking at each of the boundaries in slightly more

detail, **biodiversity loss** is principally a matter of changing land use away from more natural systems to managed systems and agriculture – the clearing of forests, for example. Overfishing also contributes to the problem in the oceans, and invasive species are a major cause of biodiversity loss in both land and sea ecosystems. Globally, land use change is driven disproportionately by the growth of grazing livestock production, and to a lesser extent by first generation biofuels. Our scenario does not use these first generation biofuels at all. Rather, it generates a large quantity of biomass crops that offer richer habitat possibilities than typical cropland (Haughton et al., 2009) while increasing forest area and maintaining, or in some cases restoring, habitats of ecological importance – peatlands, for example.

Water consumption becomes a global issue in terms of 'embodied' water in goods and food (Hoekstra, 2013). Our scenario at least partly addresses this question with a reduction of food imports and zero imports of water intensive livestock products (Mekonnen and Hoekstra, 2012).

The problems of **nitrogen and phosphorus** excess are also connected with grazing livestock production, since more nitrogen and other fertilisers are required to produce animal rather than plant protein – simply because of the quantity of land used (Lilywhite and Rahn, 2005). Although in our scenario we do not explicitly model fertiliser application, reducing the amount of land used to produce foodstuffs is likely to decrease the amount of fertiliser used to some degree (Sutton et al., 2013).

Although it has not been possible to investigate in detail the interaction of the Zero Carbon Britain scenario with these proposed planetary boundaries, the requirements have consistently been kept in mind. The technical choices made in our scenario aim at genuine sustainability, not merely a reduction in impact.

3.9.3 Employment

By its very scale, the transition outlined in this report holds the potential to be a powerful generator of employment – not only in emerging industries like offshore wind, but also in existing technology and manufacturing sectors: construction and transport, for example. There are also significant new employment potentials in land based industries, such as growing energy and fuel crops and carbon capture processes. Denmark and Germany have already set an example by decarbonising much more rapidly than the UK, and in the process creating employment on a very large scale (see box below).

Although some jobs will inevitably be lost in conventional energy systems, new jobs in renewable energy, construction, transport and agriculture should more than compensate, though the new jobs will be different and may not emerge in the same locations – perhaps rejuvenating rural and ex-industrial areas.

Employment will be created in powering down

An example from Germany

"Germany already has twice as many people employed in the renewables sector than in all other energy sectors combined. An estimated 387,000 jobs had been created in the renewables sector in Germany by 2011, far more than the total 182,000 people working in all other energy sectors. By 2020, more than 600,000 people are expected to work in the renewables sector – roughly as many as are currently employed in the automotive industry.
Wind, solar, biogas, and geothermal power provide employment opportunities for many traditional industries. Heavy industry also benefits in a number of other ways. For instance, wind turbine manufacturers are now the second largest purchaser of steel behind the automotive sector. A number of struggling ports in Germany are also positioning themselves for the offshore wind sector.
While some of these are manufacturing jobs, many others are in installing and maintenance. Jobs for technicians, installers, and architects have been created locally and can't be outsourced. They already have helped Germany to come through the economic and financial crisis much better than other countries."

Extract from *German Energy Transition - Arguments for a renewable energy future* (Morris and Pehnt, 2012).

energy demand through a massive national programme of retrofitting buildings, energy efficiency improvements and reshaping of our transport systems. Powering up the UK's renewable energy assets also offers significant employment, particularly if the generation equipment can be manufactured here. Further employment opportunities would be found in sustainable forestry management, the conservation sector and in the growing fields of biomass for carbon neutral fuels.

There is a clear need for further research to map out this employment potential in more detail. Many studies cite a variety of estimates, based on widely differing assumptions, which makes it difficult for an accurate analysis of the effect of our scenario on employment. However, our rough estimate of the job creation potential is as follows:

Power Up: 1.33 million jobs

The Renewable Energy Association (REA) report, *Renewable Energy: Made in Britain* (2012), estimates that the UK renewable energy sector employed 99,000 people in 2010-11 and 110,000 people in 2012. The Department of Energy and Climate Change's *Renewable Energy Roadmap 2012 Update* suggests that, in addition to the REA's estimated 110,000 jobs directly in the renewable energy sector in 2012, there were another 160,000 jobs along the supply chain.

Projecting these figures using historical growth rates in the sector, and including the impact of the UK reaching its binding European Union target of 15% of energy generated renewably by 2020, the REA estimates that around 400,000 jobs would be created – or in other words, for every 1% of our energy produced renewably, about 26,700 jobs are created.

Extending to all renewables, and extrapolating from this estimate, we might initially and tentatively conclude that to provide 100% of UK current primary energy from renewables by 2030 would require some 2.67 million jobs. However, energy production (due to decreased demand) in the scenario is estimated at around half of the current level. The number of jobs in the energy sector (and supporting services) in our scenario would therefore be approximately 1.33 million jobs.

Additionally, regarding wind and marine renewable energy deployment in the UK to 2020, the Renewable UK (RUK) report *Working for a Green Britain* (2011) stated:

"The High Scenario represents a very ambitious but achievable outcome... An overall 10-fold increase in the deployment of wind and marine technologies (51.8 GW) could support over 115,000 full time equivalent jobs, 73,000 of these would be working directly in the sector and the rest in the supply of wind and marine energy related goods and support services."

Our scenario envisages almost four times the capacity of wind and marine technologies, meaning roughly 460,000 jobs may be created in this sector. We can assume then that the remaining 870,000 jobs (out of 1.33 million) would be in solar power, geothermal, synthetic gas and liquid fuel production etc.

Power Down: 150,000 jobs

Job creation potential in energy efficiency measures is a little more difficult to quantify. It was estimated in the *ZeroCarbonBritain2030* report (Kemp and Wexler, 2010) that some 170 jobs should be created per TWh in energy saved. With about 900 TWh of energy demand reduction measures in this scenario, roughly 150,000 jobs might be created.

Land use: 40,000 jobs

There are currently about 450,000 people employed in UK agriculture, of which about 300,000 are in the livestock or mixed sector (National Careers Service, 2012). Some 40,000 are currently employed in forestry and the primary processing of wood products (Forestry Commission, 2012). In our scenario, agricultural cropland area remains about the same (though the product mix changes). The livestock sector shrinks significantly, though much of the land released is used for growing a variety of energy and fuel crops. It is hard to estimate the exact balance, but the total number of agricultural

workers would probably be much the same as today. In addition, the scenario envisages more than a doubling of forested area. We therefore estimate the creation of about 40,000 additional jobs in forestry and the primary processing of wood products.

Other employment opportunities may exist in the verification and validation of carbon capture schemes, in biochar production and in the restoration of conservation areas such as peatlands, however estimates of these figures are hard to find.

Overall, our estimate of the job creation potential is just over 1.5 million new jobs.

3.9.4 Wellbeing – measuring what matters

Wellbeing describes the health and social, economic, psychological or spiritual condition of an individual or group, and is more closely associated with 'quality of life' than 'standard of living'. In order to better understand what we really mean by wellbeing we need to be sure we measure what matters. Key indicators include our physical and mental health, our impact on the environment, education, recreation, leisure time and social belonging.

So, as we explore a scenario for moving away from

fossil fuel dependency whilst also preparing for the climate impacts already in the system, we must adopt these new indicators to chart how this influences our wellbeing – both in our personal lives and collectively as a society.

Measures of collective wellbeing

Traditional collective measures, such as Gross Domestic Product (GDP), do not offer a reliable measure of our wellbeing – in fact these might actually increase following natural disasters such as floods. Since the 1970s, the UK's GDP has doubled, but our perceived 'satisfaction with life' has hardly changed (Aked and Thompson, 2011). Such measures not only fail to register the damage we do, they also fail to actually tell us how well we are doing.

The new economics foundation's (nef) Happy Planet Index (HPI) is an example of a global measure of sustainable wellbeing (Abdallah et al., 2012). It tells us how well nations are doing in terms of supporting their inhabitants to live good lives now, while ensuring that others can do the same in the future.

Measures of individual wellbeing

The new economics foundation's (nef) *Five Ways to Wellbeing* report (Aked and Thompson, 2011) identified a set of evidence based actions to improve wellbeing:

- Connect with your friends, colleagues or local community.
- Be active, walk or run, step outside, cycle, play a game, garden or dance.
- Take notice, be curious, catch sight of the beautiful and remark on the unusual.
- Keep learning, try something new, set a challenge you will enjoy achieving.
- Give, do something for a friend or a stranger, thank someone, volunteer.

Zero Carbon Britain and wellbeing

Any decarbonisation framework will, of course, require a new approach to many of our current lifestyle choices. The trick is to find synergies between the changes required to reduce our emissions and the changes that can increase our wellbeing.

Whilst measuring the impact of our Zero Carbon Britain scenario is challenging, we can use it to begin to explore how a decarbonised society might affect our wellbeing.

For example, our scenario includes challenging consumerism, thereby increasing resilience – both of our environment and our society, by changing our diet, increasing levels of physical activity and reprioritising how we spend our time. There will be more room for natural spaces around us, and more people working closer to nature, and perhaps closer to home. All of which hold the potential to deliver direct benefits to our wellbeing.

By pursuing real needs over induced wants, and through finding ways of defining ourselves and our relationships independent of material possessions, we can face up to our fossil fuel addiction and decarbonise our diet, buildings, energy, travel, water, work, clothing, heating and holidays. Rising to this challenge may offer us an additional benefit by way of a rich sense of individual meaning and collective purpose that is perhaps lacking in today's society.

3.10 Other scenarios

Mykl Roventine / Creative Commons 2.0 Generic

Our scenario describes *one* way of 'getting to zero'. We modelled a technically feasible scenario that could be implemented immediately, given the political and social will. It doesn't depend on techno-fixes promised in the future, and illustrates that we can cater for our own energy and nutritional needs whilst taking responsibility for our GHG emissions and playing our part in a global transition to a safer world.

Along the way, however, we have had to make some compromises. Mostly we think the changes we propose would in fact have multiple benefits (see 3.9 *Benefits beyond carbon*). But there are some things that may be less palatable to a lot of people – eating much less meat and flying much less frequently, for example.

However, there are many other zero carbon scenarios. Here we explore a few of the alternative options. We have not modelled these explicitly and so can't be certain of how (or if) they add up, but it is interesting to look at where the options open to us might lead, and what other kinds of futures they create.

3.10.1 Scenario variations using ZCB rules

Even within the rules we set ourselves to create our scenario (see *3.1 About our scenario*), there were options and we had some choices to make. Here we discuss what alternative choices could be like.

Different ways of eliminating emissions from energy

Most scenarios concentrate on energy because it makes the largest contribution to our GHG emissions (around 82% in 2010 (DECC, 2013)). There are very many ways to reduce emissions from the energy sector – David Mackay provides examples in *Sustainable Energy Without The Hot Air* (Mackay, 2009).

In a similar manner we can construct widely different energy mixes that equally serve to deliver a zero carbon supply, and then explain the various choices.

Virtually all analysts agree that a standard mix of renewables will be viable by 2050, if not well beforehand (since we, and many other countries are already generating power using renewables). Any scenario is likely to include biomass of various kinds, hydropower, solar, wave, tidal and ambient heat for heat pumps (these are examples of what might be included in 'other UK generated renewables' in figure 3.40), and a generous wind component (both onshore and offshore). Equally, any scenario would need a reorganised electricity grid. These are common factors in almost all scenarios (Wiseman and Edwards, 2012).

Beyond this there can be substantial differences. In figure 3.40, **Scenario 1** illustrates a kind of mix that features in many conventional zero carbon energy scenarios. Similar to our energy system today, the focus is on supply rather than demand (small amounts of demand reduction meaning a large energy supply is still required), including a fairly high proportion of baseload sources. It relies heavily

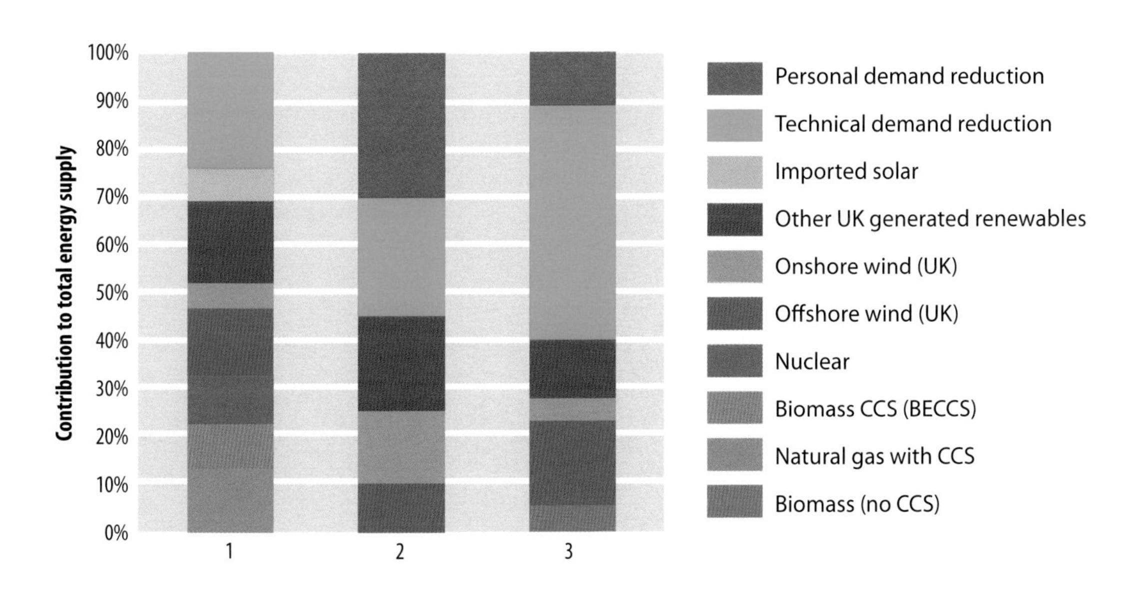

Figure 3.40: Illustrative examples of three different zero carbon energy scenarios. Note these are not calculated, but are illustrations of a concept – the specific percentage contributions of various measures to the energy supply may be different to those presented here.

© Patrick Laverty

on nuclear power and biomass energy, plus imports, perhaps via a European super grid. Bending our rules slightly, it uses natural gas with carbon capture and storage (CCS), but because this is not strictly carbon neutral (there are still some emissions from burning the fossil fuel that are not captured (DECC, 2012)), it requires a proportion of biomass with CCS (Bio-energy with Carbon Capture and Storage – BECCS) to provide carbon capture of the remainder of emissions from gas power production.

In complete contrast is **Scenario 2**, which emphasises demand management rather than supply, relying partly on consumers to reduce energy consumption through lifestyle changes – 'personal demand reduction'. It uses only renewables to provide energy, with a particularly large onshore wind component (though again, this energy mix could be vastly different). Within this scenario there could also be a substantial micro-generation element, and possibly decentralised management (community wind turbines or solar farms, for example), resulting in big differences in the types of energy generation in different parts of the country. Occasional shortages and fluctuations in supply might be accepted as a reasonable exchange for low cost and minimal environmental impact.

Scenario 3 is most similar to our ZCB scenario energy mix, with a high level of attention to 'technical demand reduction' (measures like insulating homes and using efficient appliances) and therefore fewer requirements for lifestyle changes. The energy inputs are again all renewable, but provide generous amounts of back up, and an important role for hydrogen (with biomass) in balancing supply and demand (for details on this, see *3.4.2 Balancing supply and demand*). It would be a high-tech and centrally managed system much as we have today, but with more emphasis on demand-side management (see *3.3 Power Down*).

Different ways of reducing other emissions

Any of these alternative mixes would mean emissions associated with energy production would be reduced to zero. But there would still be about 18% of emissions remaining in the scenario (DECC, 2013).

To get to zero, a typical approach reduces these non-energy emissions as far as possible, then balances what remains using a variety of processes that capture carbon. Both these steps can be done in a variety of ways, giving rise to many possible scenarios.

The ZCB approach is to apply technical measures wherever practical. These apply mainly to non-energy industrial, household and business emissions, as well as to those from waste (see 3.5 *Non-energy emissions*).

A major decision regarding emission sources that do not have convenient 'techno-fixes' is: to what extent are we willing to change our lifestyles? It is more difficult to decarbonise without social and personal choices and trade-offs. For example, as mentioned in 3.6.1 *Agriculture, food and diets,* the best way of reducing emissions from agriculture would be to eliminate meat and dairy products entirely from our diets. Since GHG emissions from flying have an amplified effect higher in the atmosphere, not flying at all would eliminate this component. If we want to eat a bit of meat or dairy, or we want to fly a bit, we have to capture the exact carbon equivalent to achieve net zero emissions.

Fuelling transport (aeroplanes and heavy commercial vehicles, for example) using fuels derived from biomass; feeding ourselves adequately; providing a portion of energy for heating, industrial processes and back up; and capturing any carbon that is still emitted by any of these non-energy processes, all require land, which is limited.

The limitation arises from the need to respect the global context: to minimise claims on overseas land that others might need for their own decarbonisation process.

Unfortunately, there is not enough land in the UK to do everything we are used to doing and still meet the carbon budget. Almost 80% of land in the UK is currently dedicated to food production (the majority of which is used to graze livestock), and only 8% is not currently managed or productive in some way. This 8% likely contains some conservation areas and particular habitats that are rare or protected in some way. In short, there is little space for growing aviation fuel, and not nearly enough space to balance out the current emissions associated with agriculture and/or flying using processes which capture carbon – planting new forests or restoring peatland, for example (see 3.6.3 *Capturing carbon* for more detail).

Therefore, an unavoidable change is to relinquish some of the grassland currently used for grazing for other uses. There are trade-offs to be made, for example, between flying and eating meat and/or dairy products – both of which contribute to climate change and take up land. Generally, we find that more of one means less of the other.

To remain zero carbon, each component of land use that still ultimately leads to GHGs emissions (growing biomass for aviation fuel, or grazing livestock) must have a complementary area dedicated to capturing the carbon it emits. We must also be careful not to release carbon from soils and plants when changing how we use land – we have to match our demands on land with the type of land available. In these ways, there are limits on how much of certain activities any scenario can contain and still remain zero carbon. In short, we have to make compromises, and perhaps prioritise lifestyle choices.

In our scenario we provide a balanced, abundant diet for the UK population (but with much less beef, lamb and dairy products); sufficient energy for heating and energy system back up; and sufficient fuel for most of today's transport needs aside from aviation, where we only have enough land to provide for a third of today's international flights. We also double the forested area in the UK and restore peatlands to capture carbon with the added benefits of increased biodiversity and more 'natural spaces' to enjoy.

3.10.2 Breaking the ZCB rules

There are, of course, many other scenarios that could be constructed, by changing the rules by which we play the game. We could, for example, include more technical fixes currently in research or early developmental stages, which would in many cases reduce those last few emissions further and would alleviate some of the demands on land. We've highlighted some promising technologies throughout the report, but have not included them in our scenario.

Furthermore, we could depend on international connections for energy provision – either balancing supply and demand via importing renewable electricity from Europe, or importing fossil fuels and coupling them with CCS and BECCS technologies. This would also reduce demand on land. With these types of changes, it might be possible to keep levels of meat consumption or flying closer to what they are today.

To balance the extra emissions, we could use various forms of geoengineering currently in research and development, such as air capture of CO_2 ('scrubbing'), or store the gas in old, now empty, gas or oil fields. Or we could buy international credits to pay for our remaining emissions – funding the transition to zero carbon economies in less developed nations by paying so that we can emit more than our 'fair share' of GHGs, or paying them to capture equivalent carbon on our behalf.

Overall, however, most of these scenarios involve more speculative technical measures, which may not deliver on time; or they rely on resources elsewhere, of which we could easily take more than our 'fair share.'

3.10.3 Carbon omissions

It is widely assumed that decarbonisation is basically an energy problem. From a world perspective it is true that GHG emissions arise principally from burning fossil fuels but, from a national point of view, direct energy emissions might account for little more than half the total depending on what we define as 'our emissions' – meaning those we are responsible for. Table 3.5 shows the effects on the total GHG emissions of the UK in 2010 by adopting various 'frames' of responsibility.

2010 UK emissions ($MtCO_2e$)	Frame
493	Emissions from direct UK energy use.
588	All GHG emissions arising from UK territory, less carbon captured by soils and plants. Often called a 'production account' it is the basis of current international agreements on climate change (the UNFCCC's Kyoto Protocol) and official emissions targets and carbon budgets.
628	***All production GHG emissions, plus those from international aviation and shipping.***
824	Emissions associated with all goods and services consumed, including imports, minus exports. Often called a 'consumption account' or our 'carbon footprint.'
(up to) 924	All consumption emissions plus emissions associated with land use change abroad attributable to UK food consumption, sometimes referred to as 'indirect land use change'.

Table 3.5: UK GHG emissions associated with various frames, and details of what the frames include. The frame used for our scenario is highlighted in italics. Data is taken from DECC (2012), DECC (2013) and Audsley et al. (2009).

Which are the *real* UK emissions? There are good and bad reasons for choosing any of these frames but, broadly speaking, decarbonisation gets harder, and more expensive, as you move down the list. That is one reason why governments and most research institutions try to stick to the 'easy end' and assume that the rest will somehow be dealt with elsewhere. But these emissions do occur, and the responsibility has to be picked up somewhere. They are in fact 'carbon omissions' that need to be accounted for if we are to take the mitigation process seriously.

In our scenario we have adopted a compromise frame, incorporating traditional 'production accounts' and international aviation and shipping, but not imports of goods and materials, or land use change abroad that would be attributed to our food consumption.

Land use change abroad

In some accounts, land use change abroad that is attributable to food consumption in the UK amounts to as much as 100 $MtCO_2e$ per year, though our knowledge about the extent of this issue is incomplete. It is a very complex issue, but it is estimated that the problem arises largely from consumption of livestock products within a globalised market – for example, clearing forests to rear cattle that we import and eat, or to grow feed for UK livestock (Audsley et al., 2009).

For this reason, the dietary changes and food importing rules in our scenario – no imports of livestock or feed – can be considered to reduce indirect land use change effects to a negligible level.

The 'stuff' we import

In the ZCB scenario, carbon emissions from imported goods are considered only by stating that our scenario must be part of a concerted global effort to reduce GHG emissions – the UK alone cannot 'solve' climate change. Other nations also have to decarbonise at rates and along trajectories coherent with their fair share of the global carbon budget (see *2.3.1 Our carbon budget*). This means that GHG emissions associated with the production of goods that we import are accounted for globally.

However, it has been widely argued that allocation of responsibility for GHG emissions should not be on the basis of production, but *consumption* (Helm et al., 2007; Druckman and Jackson, 2009). In other words, the emissions from all goods and services should be allocated according to *who consumes them* and not from where they are produced. This is bad news for wealthy countries like us that import a great deal of goods and commodities, but good news for countries that export large amounts, like China.

Of course, whatever the accounting conventions, the total world emissions remain the same – the national totals would just be allocated differently. It could be asked then, if production accounts are good enough for current international agreements regarding emissions reduction, like the Kyoto Protocol, why quibble? The argument, however, is that accounting based on production seems somehow unfair, open to abuse, and leaves a distinct impression of accounting fraud.

For example, on a consumption basis, taking net imports into account, we find instead that not only does the UK have much higher emissions, but also that these have grown. Rather than declining from 778 $MtCO_2e$ in 1990 to 588 $MtCO_2e$ in 2010 as the headlines tell us, they have in fact increased over the same period – from somewhere in the region of 650 $MtCO_2e$ per year to about 824 $MtCO_2e$ per year (DECC, 2012).

Since fairness is likely to be a key component of any international decarbonisation process, this is an important debate (Wei et al., 2012).

Although we have not modelled it in our scenario, we can make some general suggestions about what we could do to decrease these emissions if we were to include our responsibility for imported goods and still aim for zero carbon. For instance, the emissions associated with the import of food could be reduced from a potential 59 $MtCO_2e$ (Holding et al.) to less than 1 $MtCO_2e$ (assuming a decarbonised energy and transport system), which shows what can be done through a combination of reduction in demand, altered product choice, and increased domestic production (*3.6.1 Agriculture, food and diets*).

Based on this example, a number of additional

things could help us decrease the consumption emissions from the 'stuff' we import:

- Reducing how much we buy (or consume), whether it is produced at home or abroad.
- Encouraging long-life products, product-service systems, and much higher levels of reuse and repair. This would also reduce the demand for goods.
- Importing items with lower or zero GHG emissions, including alternative low or zero carbon materials – for example, bioplastics and composites.
- Increasing imports that would constitute additional 'carbon capture' – for example, the import and use of wood products. According to our rough calculations, current use of imported timber and wood products results in an additional 42 $MtCO_2e$ captured per year (see *3.6.3 Capturing carbon*). With more use of plant-based products in buildings and infrastructure, this could go part way to 'balancing out' additional emissions from imports.
- Producing more in the UK – domestic production that the UK is entirely capable of but has systematically off-shored because production is cheaper elsewhere could be reclaimed and increased once again. This might mean higher industrial energy demand, and perhaps more non-energy emissions. We might need to install more energy infrastructure and capture more carbon as a result. Fewer imports would, however, decrease fuel demand for aviation, shipping and UK distribution even further.

Having said this, with a somewhat de-industrialised economy deeply dependent on imports for finished goods and raw materials, rapidly increasing domestic production may be problematic for the UK. Furthermore, with higher emissions at the start of the decarbonisation process, we might fail to keep to a carbon budget that would give a reasonable chance of avoiding a 2°C global average temperature rise. The purchase of international credits might be necessary to aid the transition, or a re-assessment of geoengineering options to remove CO_2 from the atmosphere may indeed have to be considered. Neither of these options, however, provide an alternative to decarbonisation – they would simply 'buy us time'.

Using consumption accounting methods would almost certainly make it more challenging to get to net zero, but some of the changes mentioned here might be beneficial to the UK – for example, we might create more jobs by producing more at home.

There are many unanswered questions, and unlike the rest of our scenario, we have not quantified any of these effects or explored the possibilities. How much more energy infrastructure would we need? What are the options for low or zero carbon materials currently? Do we have enough land to capture sufficient carbon? How much might demand for goods reduce?

These are areas we would love to look into in more depth, and will form important subjects for further research.

Chapter 4

Using ZCB

Now that you've read our Zero Carbon Britain scenario you've already begun to get to grips with the scale and seriousness of the challenges ahead. This is an important first step! Hard though it may be, recognising the magnitude of our predicament forms the cornerstone of our response.

However, there are many ways to take the next step – should we focus on 'campaigning for change' or 'being the change' ourselves?

There are many things that need to be changed at national and international levels, but if you feel starting at home is what you want to do – that's OK too.

In fact, there are many ways of being part of a transition to a zero carbon Britain, with lots of community and domestic scale choices and a wide range of campaigns and organisations to join and make our voices heard.

4.1 Changing how we think about human beings and energy

© Joanna Wright

Humanity's relationship with the incredible amounts of ancient sunlight stored up as energy in fossil fuels has brought us into spectacular times.

On the one hand, we have seen incredible advances in technology, medicine, art, science, education and entertainment. In the developed West, life expectancy has increased dramatically and many new medicines are tackling killer diseases. If you have the means, you can have most of the things you could ever want. You can listen to a perfect digital reproduction of traditional Tibetan flute music, whilst watching widescreen 3D images of the beauties of the lower Nile, and eating authentic Chilean cuisine with fresh New Zealand kiwi fruit to follow. If you wish, you can even go there and experience it all first-hand. It is a feast that we all, to a greater or lesser degree, participate in.

On the other hand, the incredible power of fossil fuels has allowed us to manipulate the world as never before. From one day to the next, we must live with, or bury, the psychological and emotional pain of the destruction, corruption, exploitation, globalisation and capitalisation of our ecosphere. Many carry this sadness in quiet solitude, often unconsciously, through life. But as the eyes and ears of the media reach out we experience, as it happens, the destruction of the planet – a spectacle we all, to a greater or lesser degree, also participate in.

Living with, and trying to reconcile, this paradox is a real problem, leaving many paralysed and confused. The destruction of our life support system is one of the most pervasive sources of anxiety of our time. Environmental groups initially assumed that we don't change our ways because we simply lack information – we don't understand the problems enough to make sensible decisions and do something about them. Experience suggests, however, that most of our numbness and apathy do not stem from ignorance of the facts, or even indifference. We are held fast, sleepwalking through the shopping malls, paralysed and overloaded from the continuous barrage of information we receive. It is estimated that the average American is exposed to more than 3,000 marketing messages every day (Futerra, 2005). As our understanding of the global energy and environmental crisis spreads, we find we have become trapped by our dependence on it and so are inevitably obliged to conform.

Although humanity's present day fossil fuel driven frenzy of production and consumption is affecting us deeply, society has created taboos against the public expression of the associated emotion and anguish. Although most of us are only too aware of the destruction of the ecosystem, we simply put it in that locker just out of our conscious thought, where smokers keep the knowledge about lung cancer or where heavy drinkers keep their awareness of liver disease.

We see the crisis, we have the solutions – but our almost total failure to take the actions that could avert it is making it increasingly obvious that our entire culture, indeed our entire civilization, is locked into denial. Denial is the primary psychological symptom of addiction. It is both automatic and unconscious. In psychological terms, denial is a 'defence mechanism'. It defends the individual or collective consciousness from some truth that they cannot afford to acknowledge because it would expose overwhelming feelings of fear, shame or confusion. As long as we remain in denial about climate change, peak oil, ecological collapse or the suffering of the majority world we are free from the associated pain, and can lose ourselves in our day-to-day lives. Yet if we do not deal with these feelings they will manifest as problems in our physical or mental condition. Over the past couple of decades, these collective fears have already transformed the way contemporary culture portrays our future: from an exciting new world of progress, where we all want to go, to a dark, dystopian world of ecological collapse.

Our relationship with energy is very powerful, shaping how we see ourselves and how we relate to the world around us; we must rethink it if we are to transform our fears into empowerment ready for the difficult collective transition ahead. By focusing on the realities of what science demands rather than simply what is politically or socially palatable, and acknowledging the UK's historical responsibility as a long-industrialised nation, Zero Carbon Britain aims to open a new chapter in the story of human beings and energy, one in which we may once again talk excitedly about the future.

4.2 Taking action in our homes, communities and places of work

There are many actions we can take at the domestic or community level, which in many cases have a two-tiered affect. Not only can we directly reduce our greenhouse emissions, but we can also begin to change how we relate to energy, personally and collectively. By pioneering real life projects we demonstrate that we have both the will and the technology for change, which can go a long way to strengthening our calls for a radical shift in policy. Better policies, in turn, should make it easier to scale up and roll out similar projects across society. We must actively explore how practical, real life changes on a local or community scale can synergise with policy actions at national or international levels, to accelerate an evolution in our relationship with energy.

'Being the change' – mapping your way with *The Home Energy Handbook*

As we set out to pioneer a path to a zero carbon Britain in our homes, communities and places of work, it is useful to explore how to 'do the numbers' – in other words, to work out where we are starting from, so we can assess our progress in cutting down our carbon emissions. To help you do this, the Centre for Alternative Technology (CAT) has produced a book based on our Zero Carbon Britain research, called *The Home Energy Handbook* (Shepherd et al., 2012). Exploring your best way forward will depend on your individual location and circumstance, but there are some common approaches:

- Get informed.
- Get a group.
- Make a plan.
- Get skilled.
- Get connected.
- Minimise demand.
- Rethink supply.
- Recycle the savings into your next action.
- Share your experiences honestly with others.

One of the best ways to begin is to get to grips with mapping your energy use. Get your hands on your data and begin to understand the scale and speed at which the different types of energy flow through your life. This process can begin with a list of the types of energy (natural gas, electrical) that you use in a typical week, month or year, and for what purpose you use it. You can find out how much of each type you use and how much the costs have increased over recent years. You can do this as a group, family or on your own; the data is there in electricity and gas bills, petrol receipts and so on. Many of the new utility bills show your consumption to date in the form of a graph.

It has been shown that simply being aware or keeping record of our energy consumption generally means we use less.

It is also worth doing a quick 'energy vulnerability' analysis for your current lifestyle. What would happen to your personal choices if any of the forms of energy you currently use became very much more expensive, or even intermittent? Assembling this picture is the first step to getting rid of that subconscious, outdated 1950s approach to energy, equipping you for the process of rationalising your energy demand and addressing your greenhouse gas (GHG) emissions.

Aside from changing your energy use, there are a few other things that you can start doing yourself to help reduce your GHG emissions that have come out of our research. As general guidelines, we'd recommend:

- Eating less meat and dairy. Becoming vegan can be challenging, but just reducing your meat and dairy consumption can have a big impact.
- Try to walk or cycle where possible rather than taking the car on short journeys.
- Support and use public transport where possible. Share lifts or join a car share scheme. Invest in an electric car if you can.
- Use natural building materials and buy wooden furniture (rather than plastic or metal – these all use energy and produce GHGs in production), and make sure all the products you use (down to the paper you write on) are sourced from sustainable forests.
- Buy things to last, or things made from recycled materials. Reuse, reclaim, recover or mend any items or materials that you can – don't just throw things away! Think about what you are throwing away and try to reduce your waste as much as possible.
- Compost your food waste and recycle as much as you can if you do have to throw things away.
- Don't add peat to your soils – use compost instead.
- Learn more about the natural world, and spend more time outdoors – it's nice, and pretty amazing once you start thinking about what it does for us without us even asking!

We know that doing some of these things are challenging, and that there are restrictions to how much we can do as individuals. This is where joining with others can be beneficial – see using ZCB as a way of 'influencing policy' below.

Please don't fool yourself – although they are important, your individual preparations are unlikely to be enough on their own. The only way ahead is to get through the climate challenge collectively, both nationally and internationally – but being personally aware and acting as individuals and communities is an important part of the bigger picture.

In pioneering real change in our own lives and in sharing our collective achievements we play a part in breaking the dangerous deadlock of 'politics as usual', and we can help lever vital international agreements by demonstrating that we have both the technologies and the will for change.

Influencing policy 4.3

CAT's Zero Carbon Britain project has offered an evolving set of scenarios that can be useful for opening the debate required for the transitions ahead. The work to date has generated significant interest, through media and events in the UK and overseas. We have presented the findings at United Nations climate conferences, in Parliament and in person to key policymakers. But it is one thing to put a copy of a report in the hand of a government minister, and quite another to build enough of a consensus across a democratic society to enable him or her to act on it. Arriving at that kind of consensus will require many people working together, spreading awareness more widely.

Certain events have been incredibly effective in building awareness. In December 2010, the UK National Climate March took our last report – *ZeroCarbonBritain2030* – as its focus, with the goals of encouraging engagement by policymakers, and of promoting the concept of rapid decarbonisation in advance of the UN climate negotiations in Cancun. More than two thousand people joined the march, many of whom have since embraced the concept of a zero carbon Britain through their own campaign organisations and regional groups.

The Welsh Government funded CAT to organise

CLIMATE EMERGENCY
Zero
Carbon
by 2030
2
30
WWW.CAMPAIGNCC.ORG
GENCY
WWW.CAMPAIGNCC.

a 'Convergence on Zero' conference in Washington in 2009 to stimulate debate in the USA. In collaboration with the International Network for Sustainable Energy, we also arranged presentations at the UN climate conferences, the French National Assembly in Paris and the Institute of Engineers in Barcelona.

There are a lot of people trying to influence policy

But despite the growing evidence of our climate and energy challenges, and a long series of international UN summits, there appears to be a block in actual decisions that could set in motion the solutions. The problem seems to be built into the very nature of a capitalist democracy. It perhaps works something like this: under our free market capitalist system, the corporations, their boards and Chief Executive Officers (CEOs), are legally obliged to maximise return to shareholders in the short-term. Any CEO who does not adhere to this will quickly find him- or herself out of a job. The modern corporate entity is compelled, by the system within which it operates, to elevate financial interests above all others. Corporate social responsibility, though sometimes yielding token positive results, most often serves to mask the corporation's required behaviour not to change its primary approach.

A multinational oil, coal or gas corporation's necessary self-interest forces it to oppose climate related legislation, not necessarily through malice, but merely because shareholder capital has been deployed in existing pipelines, mines, oil wells or refineries, and it is the legal obligation of the CEO and board to maximise return on these investments. As energy companies are very large and profitable, significant financial resources are available to be spent to this end. Many are now urging for the restoration of the 'corporation's' original purpose, to serve the public interest, and there are calls to re-establish democratic control over these institutions.

Encouraging policy change with ZCB

Gathering enough momentum to catalyse a consensus for action across a democratic society requires a wide range of people and organisations. We need your support, and there are several things you can do. You can, of course, take action by speaking or writing to your Member of Parliament (MP), Member of Scottish Parliament (MSP), Assembly Member (AM), Members of the Legislative Assembly (MLA), or Member of European Parliament (MEP) about the need for rapid decarbonisation in the UK, or specific issues regarding transport, energy provision or agriculture, that form part of a zero carbon future. The ZCB project can provide context for many issues and to demonstrate the feasibility of a zero carbon Britain (or UK!). You can also get involved by raising awareness of the project through organisations and groups whom you feel are relevant or might be interested.

There are already many campaigns working to bring about the various changes described in Zero Carbon Britain. Joining them, or using ZCB to build links between them (in order to campaign towards a common goal) is an effective way to become involved. You can find some examples of these organisations on our website
www.zerocarbonbritain.org

We'll be updating it to keep people connected.

With globalisation and the widening of access to social networking, the power of our collective creativity is amplified. Indeed, online culture is becoming as powerful as the military or politics. It's no longer just a matter of whose army wins but whose story wins! Creative ways of influencing and motivating others are constantly emerging, as witnessed in the Arab Spring, the Occupy movement and UK Uncut, among others. There are countless exciting ways to help change how we think about our relationship with energy and our GHG emissions, whether you use Zero Carbon Britain or other tools.

4.4 Zero carbon education

CAT education officers use the Zero Carbon Britain project as a teaching tool with pupils, students, teachers and tutors of all subjects.

We help learners develop skills, values and critical thinking, and we give them time to reflect, question and debate. Without accurate information, there is no foundation on which to build the well informed decisions towards which critical thinking leads. It is therefore important that our approach is backed by research, which is where the Zero Carbon Britain scenario comes in.

In 2009, the Cambridge Primary Review noted that:

"...pessimism turned to hope when witnesses felt that they had the power to act. Thus the children who were most confident that climate change might not overwhelm them were those whose schools had decided to replace unfocused fear with factual information and practical strategies for energy reduction and sustainability."

Successful education for sustainability combines visions for a positive future with accurate scientific literacy and the skills to make the vision a reality. This also makes Zero Carbon Britain a good education tool.

Anyone who has struggled to keep their attention from drifting during a presentation they had wanted to learn from will appreciate the value of good teaching techniques. We learn by doing and thinking as much as by listening. Good education can deliver difficult concepts by allowing learners to engage with the material on their own terms, or by altering the language and approach used.

Teaching Zero Carbon Britain

CAT education officers include elements of the Zero Carbon Britain project and scenario in almost every guided tour or workshop they deliver. They have also developed an activity specifically to communicate the Zero Carbon Britain scenario. This workshop has been delivered to hundreds of pupils, students,

Zero Carbon Britain workshop

Scene setting

Learners are given information about climate change and other global challenges we face. They are provided with the 'context' of the challenge and of current political targets to reduce CO_2 emissions in the UK. A large map of the British Isles is provided to prompt thinking about available land resources.

Group work

Learners divide into groups representing various 'government departments' – for example, agriculture, energy, buildings and transport – and are tasked with developing a zero carbon plan for their sector. They decide how to present their ideas to the rest of the group – with modelling clay, for instance. Each group has a pack of relevant background information – maps showing the average wind speed over various parts of the UK (onshore and offshore), where the locations suitable for tidal power might be, or for UK car use, or the impacts of the food we eat.

Sharing plans

Groups take turns to present their plans to their peers. An education officer facilitates the presentation, encouraging questions and challenges from those listening. Debate ensues and groups can adapt their plans to accommodate new ideas or challenges that have arisen.

Summing up

The education officer draws out the main points and conclusions that have arisen, highlighting common ground and challenges. The group discusses whether their own vision for a zero carbon Britain is technically possible and desirable.

Sharing the ZCB scenario

The education officer gives a brief overview of the Zero Carbon Britain scenario, drawing attention to any similarities or differences with the plan made by the group. There is time for questions and conclusions as to the feasibility and desirability of the Zero Carbon Britain scenario, and discussion relating to how it might affect people's lives.

undergraduates, postgraduates, teachers and tutors – any group of learners over the age of 12, and it is always received with enthusiasm.

The Zero Carbon Britain workshop follows principles that have proved successful in other CAT education activities:

- It allows learners to understand and come to terms with the reality of climate change, our relationship with energy and the subsequent impacts on economy, environment and society.
- It allows learners to develop future scenarios of their own using discussion and practical resources.
- It provides accurate information on which learners can base their decisions.
- It demonstrates connections between our own actions and environment and societies around the world.
- It takes quality of life into account.
- It is a practical activity that includes debate, humour and creativity.

A shared vision

From an educator's perspective, teaching about Zero Carbon Britain in this way has advantages. It allows us to understand what our pupils know and feel about issues such as climate change. It allows us to find out what pupils already understand about potential solutions, and where they still need support. This varies between groups – for example, some groups believe that nuclear power and wind energy are the only non-fossil fuel options, though they are not necessarily happy about this. Teachers then know what knowledge gaps they can fill – for instance, by teaching about anaerobic digestion, or biomass.

It is striking how similar the learners' scenarios often are to the Zero Carbon Britain scenario. Their plans combine new technology with old skills and knowledge. They include a blend of legislation and incentives. They realise that a zero carbon Britain is technically achievable and also desirable in terms of quality of life. In this way, education creates a receptive audience, eager for the vision to become a reality.

4.5 Developing a zero carbon project

There are many different zero carbon Britains that can be created (as we discuss in *3.10 Other scenarios*), but there is also huge potential – and need – for zero carbon Bavarias, Belgiums and Brazils for example. Why not start a new project?

Kick-starting your project

Many countries currently have no emissions reduction targets, so exploring zero carbon scenarios can help inform new development models that can offer an economically viable and secure energy future. The framing of any research should be carefully chosen to reflect the needs and culture of the particular country, thereby linking global carbon issues to important local concerns. This helps by embedding any research work around the issues and language relevant in the locality.

However, starting a zero carbon research programme from scratch involves a great deal of detailed work, so it may be worth considering seeking some initial kick-start funding (3 to 6 months) for a 'project development officer' to achieve the following:

- Identify relevant research in this field – other scenarios.
- Identify and engage relevant collaborators – universities, industry, non-governmental organisations (NGOs) and think tanks, for example.
- Develop a research and communications strategy.
- Investigate the required roles and write job descriptions.
- Identify key funding agencies.
- Write and submit further funding proposals.

An initial request to support a 'project development officer' offers your potential funders the opportunity to engage with your project, but with a lower initial risk. It also gives the time and resources needed to be sure the main project is well thought out and adequately resourced.

Research

Here is some advice on research methodology:

Understanding synthesis research

The development of a zero carbon scenario can be achieved through a process of synthesis or secondary research, involving the integration of a wide range of existing work. Key elements include:

- Wide and detailed investigation into relevant reports, previous research, industry and academic journals.
- Ensuring all data is robust, verifiable, compatible and reliable.
- Full citation of original sources and references.
- Clarity about the assumptions underlying your scenario.

Establish working groups

Research working groups compile the latest findings from each area and, through the work of a research co-ordinator, integrate their findings with those of the other working groups via a core model.

Expert seminars

Expert seminars can bring together, in a convenient central location, a selection of leaders in their field. It is useful to have a high profile partner organisation to make the invitations, plus an independent facilitator. Engaging with key players through the research seminars can also help with your communications strategy.

The core model

Serving to integrate data from all aspects of the research, the model allows the team to construct alternative paths to decarbonisation by varying constraints, assumptions, demand patterns, energy inputs and land use options, eventually resulting in an emerging 'favoured scenario'. It may also prove effective for verifying the model to begin by using it to represent the existing system. The model can be developed in modules that reflect the findings of the different working groups, for example, transport, food and energy supply. This model works best on an annual timescale (looking, for example, at GHG emissions per year), but as the project progresses you might find it useful to model smaller time frames – see 'Dealing with variability' below.

Choosing the software for your model is dependent upon the scale and scope of your research project and the skills and funding available. In its simplest form, the model is an accountancy tool, constraining the scenario to a defined carbon budget over the chosen transition period, and enabling a balancing of the books for the supply and demand of energy. Though energy modelling software is available commercially or from research institutions, national carbon models specifically designed for this type of project are hard to come by, making it likely that your research team will actually make the best one, encompassing your approach to the problem and based on the data you have available to you. This isn't as daunting as it sounds!

Accessing data

To ensure the model is robust and its results verifiable, input data must be carefully selected. There are benefits to scaling up data from real life renewable projects rather than only using theoretical predictions. Good data sources include government, industry, energy think tanks and academia, but much of this may be sensitive due to it being 'commercial in confidence'. A number of input sources are now being used that were not available only a few years ago. In the UK, for example, current national breakdown of energy consumption is derived from the government's 'Digest of UK Energy Statistics' (DUKES).

Dealing with variability

Nobody seriously questions the fact that renewable sources like offshore wind can produce a huge amount of energy. However, if we are serious about proposing scenarios where most or all of our energy needs are met by renewables, then we need to be able to explain, with confidence, how supply and demand are matched at any given moment. To provide a detailed analysis of variability, your research will have to model hourly supply and demand patterns using national weather data.

Communicating your findings

Our experience to date has shown that there are benefits to following up the research phase with a communications phase. Too many good reports end up on the shelf because there is no associated programme to ensure people know about the findings. Through media publicity, liaising with networks of organisations, workshops, speeches and through presentations to universities, community groups, campaigners and policymakers, it is possible to create a significant level of support and public engagement.

More information can be found on our website **www.zerocarbonbritain.org** Get in touch if you have any questions and we'll do our best to help. Good luck!

4.6 Reclaim the future: engaging with arts and creative practice

Communicating the Zero Carbon Britain scenario means helping people visualise what it could be like to live in the year 2030 if we rose to the challenges of the 21st century. To offer a context to this, we looked at how our society currently portrays the future, and how this has changed over time. We quickly became aware that there are actually very few positive visions of a 21st century future. Dystopia and ecological collapse almost always abound when contemporary culture looks even ten or twenty years ahead. Be it a novel, a film, a TV series or a computer game, the setting is dark. From *Children of Men, The Road,* and *28 Days Later* to *The Survivors* – the list seems endless. Yet back in the fifties, sixties and seventies, the way we projected the future felt very different. The likes of *Dan Dare, Thunderbirds* and *Star Trek* were going to take us away to exciting places with transporters, hover bikes and jet packs.

As the seventies rolled into the eighties and nineties the wonders of science and technology were seen to be smashing into the limits of the planet's ecosystems. Alarm signals from the Green movement, along with Bhopal, Chernobyl and a wide range of other major catastrophes, led us into a different way of seeing our future. In film, a tipping point was perhaps *Blade Runner,* where the future became much darker.

Of course, setting any human drama in a tragic famine situation would not make palatable viewing, so a number of clever tricks are deployed. Either 98% of the population dies from 'the virus' before the film begins and the story is based around those relearning to plough with oxen in a deserted Somerset mansion – or – 98% of the population are converted to 'zombies' so that if you have to shoot a few dozen of them as you escape the city with the medicine for the sick child, no one thinks any the worse of you.

Despite the fact that a great many of us would like to explore the drama of human interaction set against a backdrop in which we are rising to our 21st century challenges – the artists, novelists, filmmakers and playwrights usually choose to paint it black.

But if society is unable to imagine a positive future, then we won't create it.

There is, therefore, a need to forge direct links between those working in the arts and sustainability to create a community of practice amongst people who understand the need to catalyse big shifts in how we think.

In tackling issues of race, gender and class, arts and creative practice have shown they can reveal our blind spots and help us see our prejudices; they can break through denial and catalyse a transformation of attitudes and behaviours.

The arts offer a much needed mirror that can help individuals and societies reflect on where we really are, and help us to explore the alternatives. Although science based reports such as this can show a way forward, when the arts and science work together we can begin to visualise what it might actually be like to live and love in a world where we are rising to the demands of the 21st century, and so reclaiming the future.

Chapter 5
ZCB and...

Artmasterrich / Creative Commons 3.0 Unreported

Achieving this Zero Carbon Britain is no small task – even just the modelling project! Our small but dedicated team of researchers has worked hard to figure out the technical and mathematical constraints on our scenario, but we can only do so much.

Over the last few months, we have been encouraging contributors to write a series of discussion papers entitled "ZCB and ..." to probe, ponder, reflect and imagine what a zero carbon Britain might be like. We asked for their help to raise awareness of a more carbon responsible society, by looking at a diverse range of impacts of a zero carbon Britain. From faith groups to farmers, from restaurants to rugby teams, the aim is to get people talking about what it would be like to live in a world where we rise to our 21st century challenges.

We had a great response – a huge thank you to everyone who has contributed their time, energy and expertise free of charge. Below is a taster of what we received – a few select pieces that give you an idea of all the exciting topics that relate to Zero Carbon Britain. But there are still many more interesting questions to ponder and discussions to be had. All of the papers, including the ones featured here, are available to read, download and share via social media on our website **www.zerocarbonbritain.org**

If you can't find what you're looking for, or would like to write one of your own, why not get in touch. Contact details can be found online.

5.1 ZCB and drivers

Richard Hebditch of the Campaign for Better Transport

Even more than with other sectors, there is a dangerous complacency amongst policymakers about reducing carbon emissions from transport. Transport is seen as difficult, the last sector to contribute its share of carbon reductions. And when it does come to reducing CO_2, the focus is on two big wins with electric vehicles and biofuels. No hard choices need to be made and the current transport mix can continue with the same levels of energy consumption and without the need for a radical rethink of how transport can be delivered. There is also the comforting thought for transport policymakers that the increase in biofuel usage and in electricity production is someone else's problem.

All this complacency might be forgiven if carbon emissions were falling from transport, but this is not the case. And it is not as if our current transport system is delivering the country's wider needs, whether framed in environmental, social or economic terms.

We can, however, have an approach that delivers these outcomes and also cuts carbon, achieving the vision of a zero carbon Britain. Key to this is to cut traffic, to cut the distance travelled by car year-on-year. Indeed, even with the most ambitious electric vehicle roll-out, the Committee on Climate Change (CCC) says that cuts are still needed to achieve the government's carbon reduction targets.

But politicians of all parties are fearful of the reaction of drivers to any policies that might affect them. The legacy of the fuel protests of 2000, the massive petition against road pricing in 2007 and the wider claims of a 'war on the motorist' blind politicians to measures that would benefit many, and ignores the real victims on our roads: pedestrians and cyclists.

But there are genuine concerns that underpin these more extreme protests and views. Cars are the default mode of transport for most and people cannot necessarily see how they can use alternatives. The cost of petrol is a major burden at a time when people's wages are under pressure, and they see prices in general rising. Plus people can feel they have to travel by car when local shops and services are closing. Any attempt to move to a more sustainable transport system has to recognise these things.

The scale of ambition needed to cut net carbon emissions to zero by 2030 can only be delivered upon if we start with where people are now, listen to their concerns, and build an approach that works with them.

Firstly, we need to consider that there are some significant changes in car ownership and use. Where public transport is good and there are local shops and services (otherwise known as London) car use is falling fast, with car use by Londoners declining by 35% over the past 15 years. Even outside of London, car use is falling. And younger people are changing their travel behaviour, with both young men and women less likely to have a driving license – for 17-20 year old men declining from 52% in 1991 to just 31% now. This is partly driven by financial pressures, but car manufacturers are also worried by the decline in cars as objects of desire, as young people place a greater priority on online networking (which also drives changes in travel behaviour).

Secondly, we know that motorists are not all the same and have quite different views on their use of cars. The transport academic Jillian Anable has done work to segment the population based on their attitudes to driving. Her research suggests that less than one in five is a 'die hard motorist'.

There are also strongly held attitudes that can be built upon to create a consensus for change. For instance, the public are sceptical about road building. Beyond the small circle of people around George Osborne, most think new roads will just create more traffic and solve nothing. And we know that most people would cycle more if they felt safe to do so. They also want to preserve local shops and services and to have friendly communities with green space nearby and opportunities for children to play.

With this in mind, the measures in the Zero Carbon Britain report could win popularity. To do so, we need to ensure that they are fair. For instance, carbon taxes would provide a clear signal to people to switch to lower carbon modes, but in the absence of other policies, there is a danger that those on lower incomes would end up paying for the rich to make those carbon savings (such as with subsidies for expensive electric vehicles). Carbon taxes can be fair if there are measures to compensate those on lower incomes, through higher universal credit payments and changes to income tax thresholds, for instance.

Thirdly, people do need realistic choices. The development of smart tickets like Oyster cards coupled with much better online information about services is starting to make door-to-door journeys by public transport much easier. But, outside of London, public transport is still very fragmented, with few incentives for individual bus or train operators to link services or provide simple and affordable tickets across different services. It is not enough to leave public transport provision to private companies. Often the measures that would make public transport a realistic choice don't provide a profit, and so central and local government have to step in.

Finally, the prioritisation of walking and cycling in our towns and cities has to be rooted in a conception of what we want places to be for. We need a long-term vision of these as places for people. This is the lesson of successful towns and cities for walking and cycling like Copenhagen, where a step-by-step approach has worked successfully. This agenda is not justified solely on carbon reduction terms, but on the benefits of cutting traffic in the places where people live, work and shop, as well as the benefits for individuals in more physical activity and avoiding high petrol costs.

As a start, the UK government should work with the devolved administrations to reduce the cost of travelling by public transport, rather than simply looking to expensive cuts in fuel duty. We also need national and local governments to work together to make green transport choices easier. There are good examples of this with the Local Sustainable Transport Fund in England and the Active Travel Bill in Wales. But we should be prepared to go further and ensure that the increasing powers for local councils are matched with increased responsibility to act on climate change, for example, through local carbon frameworks with clear targets.

But above all, we need politicians to recognise that whatever transport mode we use, we all are part of an interlinked transport network. The tribal language of 'ending the war on the motorist' has to stop. Travel should broaden the mind, not divide us. If we start with recognising that, then together we can ensure a transition to zero carbon transport that works for people and the planet.

About the author:

Richard Hebditch is Campaigns Director at the Campaign for Better Transport. The organisation's vision is a country where communities have affordable transport that improves quality of life and protects the environment. Achieving that vision requires substantial changes to transport policy which Campaign for Better Transport aims to achieve by providing well-researched, practical solutions that gain support from both decision-makers and the public.

5.2 ZCB and community energy

Community owned renewable energy: an agent for opinion change

Vijay Bhopal and Darcy Pimblett of the Sustainable Community Energy Network (SCENE)

A zero carbon Britain by 2030 may appear to be a wildly ambitious target but, when broken into its constituent parts, it becomes less speculative and more manageable. Here we look at one of the fastest changing of those constituent parts – electricity production. We assess the role that local ownership of renewable energy generation is playing in shaping public attitudes towards renewable energy and energy use behaviour.

Over the past two years, SCENE has carried out research and consultancy work focused on community ownership of renewables. Our organisation's core belief is that a rapid expansion of installed renewable energy capacity coupled with mass cuts in energy use is essential to achieving the emissions reductions required to avoid catastrophic climate change. However, through our work with communities, we have found that despite a present concern for global-scale issues like climate change, community renewables projects are predominantly driven by more immediate community interests, such as economy and energy security.

This is neatly shown in our research paper *A Report on Community Renewable Energy in Scotland* (Harnmeijer et al., 2012), which assesses the primary motivations of project leaders in establishing community energy projects. As displayed in the report, lowering the carbon footprint/increasing community awareness of energy issues lags far behind economic factors where motivation for community energy projects is concerned – it holds weight at just 16.8% compared to 70%.

We see five major benefits of successful community owned energy projects:

1) **Dispersal = Resilience.** Energy production in the hands of local communities creates islands of security during grid outages and contributes to voltage stability.
2) **Financial and other benefits.** Community renewable energy projects provide economic, environmental and social opportunities.
3) **Heightened energy efficiency and consciousness.** Ownership of renewable energy generation helps to promote greater energy efficiency and awareness of energy use.
4) **Ownership = Support.** Local community project ownership helps overcome public opposition facing renewable energy development in general.
5) **Market access and sectorial synergy.** Communities present an important potential source of investment, and revenue from community-led renewables projects is often recycled back into the renewables sector.

As the number of community energy projects in the UK grows, we are seeing that the 'softer' benefits, (3) and (4), are becoming evident and are beginning to have macro-scale effect. Ownership is starting to change the way people think about energy use and renewable energy development in general. This is far above and beyond the impact that the project leaders aimed for initially – that is, financial benefit (2). We have even found that community projects cause these positive secondary impacts even if they fail to reach the operational stage – stemming from the deeper understanding local people gain of how energy is generated and used.

As a consultancy, we have been involved in a range of projects with varying degrees of community support and opposition. However, what we have witnessed is that ownership of renewable energy developments has a powerful impact in changing community attitudes. In some cases, the ownership of a single turbine in a large wind farm has changed a whole town's perception of not only that wind farm but also all the other renewable energy developments in the region. A recent client of ours is looking to develop a hydropower scheme through their town.

The potential financial benefit to the community has not only changed the opinion of the community council towards the project, but has also spurred talk of energy meters and insulation projects for all.

With more than 900 community energy projects ongoing in the UK (now mapped through our SCENE Connect project), we believe that such secondary benefits are not only having an impact, but are gaining momentum and should be explored in further detail. We believe the wider attitude changes caused by community ownership of renewables will prove to be vital.

One particular area that requires more analysis is the influence that community renewable energy projects have on energy consumption behaviour at a household level. A significant body of research indicates that households that install renewable energy systems are inclined to reduce their energy consumption. This is generally the result of increased education and communication amongst individuals, which is driven by the presence of their own renewable energy system. This knowledge encourages 'soft' benefits, such as energy conservation and load shifting (changing habits in order to use more energy when the renewable resource is available).

Whether or not community owned renewable energy systems have a similar effect is unknown. However, it has been suggested that community energy organisations create a unique social environment that may influence behaviour (Devine-Wright et al., 2007), and may create "a positive social context for individual action" (Rogers et al., 2011).

SCENE/University of St Andrews researcher, Ashton Whitcomb, is currently undertaking such research, exploring the effect of a 9.3 kW community owned solar photovoltaic (PV) project on the energy use of individuals in Eskdalemuir, Scotland. Whilst this study is currently incomplete, preliminary analysis suggests that the project has led to an increase in the understanding of energy use and conservation behaviour amongst community members. As one respondent said,

"Since the installation of the PV, everyone's thinking about it more."

What is clear from our own experience is that community owned renewable energy projects can act as powerful agents for change amongst citizens, impacting communities above and beyond the bottom line aimed for initially by the majority. A zero carbon Britain requires this sea change in education and opinion change, allowing a thriving renewable energy sector coupled with a deeper understanding of energy itself. Local ownership of renewables is the perfect catalyst.

About the author:

Vijay Bhopal is Operations Director at SCENE, an Edinburgh based social enterprise which specialises in community energy research and consulting. Darcy Pimblett is a Project Coordinator at SCENE, a new arrival from Melbourne, Australia – he has a background in energy efficiency in the built environment.

5.3 ZCB and farmers

Reactions to land use change

Sophie Wynne-Jones

ZCB starts from a radical perspective of where we need to be to live in a zero carbon way, rather than concentrating on where we are now and how to incrementally change that for the better. This could be understood as devising a theoretically possible model (based on the restrictions of current science and technology) and worrying about how to get there later – having to work within, or indeed often against, the parameters of human behaviour and irrationality. Unsurprisingly, one of the major hurdles ZCB has to overcome is this human dimension. Here I consider this question in relation to the land use scenario and reflect upon some of the concerns farmers in the UK may have when faced with the imperative to decarbonise.

The land use scenario set out in ZCB suggests a major change to many aspects of current practice, including the introduction of new biomass crops, greater cover of woodland and other carbon-rich habitats, and conversion away from livestock farming to crops. The insights set out here draw on interviews with farmers to gauge their reactions to such changes in land use and the potential for incentive mechanisms, including payments for 'ecosystem service' provision (see http://www.walesruralobservatory.org.uk/our-publications).

Ecosystem goods and services are the benefits people obtain from ecosystems. They include 'goods' such as food and water, the regulatory 'services' of flood control and carbon mitigation, and the cultural and spiritual benefits of the environment. Offering payments to farmers for such goods and services is one obvious avenue towards decarbonisation – in other words, we pay farmers to manage their land to enhance carbon sequestration (for instance, payments to incentivise carbon friendly methods of food production and/or alternative land use strategies). This is called 'paying for ecosystem services' (PES), and while it has its critics, the following discussion focuses primarily on the responses from farmers to PES and measures like it.

Would such a proposal be viable? The first point to note is that farmers – like most people – do not prioritise financial gain above all other factors. For many, farming is a lifestyle choice as well as a business and so a range of complex factors come into play. For instance, social norms, self-perception and identity are key factors in the processes of decision-making and, as a consequence, we need to be aware that alternative land uses may conflict with how farmers perceive their role.

"As farmers are getting older perhaps a lot of us will see it as a bit of a pension ... But most of us, I think, want to produce food, that's the main thing we want to do..."

Equally, whilst farmers are clearly influenced by financial incentives, they are not prepared to make changes which they feel would cause irreversible changes to the land over the long-term, by reducing fertility or allowing scrub and woodland encroachment:

"It is quite hard, you have to make business decisions at the end of the day and when you are offered money to take the sheep off the hill it is very difficult to go against that ... but for how long can you make those short-term business decisions to the detriment of the long-term?"

This point of view has become increasingly pressing in light of the recent emphasis upon food security. Consequently, to engage farmers it is important that decarbonisation should be done in a way that does not create trade-offs between growing food and managing carbon. For example, planting woodland and other habitats for carbon storage should be done in consultation with farmers and in a way that works with their farming systems. In this way we are less likely to plant up their best agricultural land, or large open fields, but to better place trees and scrub in margins and on poorer quality land.

Another point that the farmers in our study highlight is the need to take account of carbon management across food supply chains, given the increasingly long distances travelled in processing, retail and consumption. Hence, farmers argued that there should be more emphasis on local procurement to reduce carbon footprints. This fits well with the ZCB scenario, but it is clear that major changes to the current patterns of processing and retail need to occur. This would involve renewed investment in the local infrastructure that was lost through previous processes of rationalisation – we no longer have the small-scale abattoirs and dairies that are essential to re-localise our food networks, but they could perhaps be reinstated.

In relation to changes in crop type and shifts from livestock to crops, it is useful to note that many farmers have only begun to specialise in livestock over the last twenty years due to economic pressures. Prior to this, most farms across the UK were mixed, and even in the most unlikely areas there are records of crops being grown. So it is possible that farmers may be more open to this aspect of ZCB. But again, it remains important to start work from existing cultural norms and expertise, and not to expect a farmer to completely convert their livestock farm to crops. Similarly, the introduction of novel crops may take longer to gain credibility simply because they are new and untested, from the farmers' perspective.

Finally, if we do require farmers to make radical alterations to their current practices, we need to explain in layman's terms why these changes are needed and how the science of climate change and carbon sequestration works. A failure to do so might mean farmers remain cynical and unconvinced, which is a particular issue in the case of the older generation of farmers who remember being encouraged to intensify food production in a way that we now realise causes a negative impact on the environment.

"I don't know much about this carbon ... nobody's come here to explain ... how does it go up to the atmosphere, does it go from the bare peat or, I don't know ... It would make a difference if we were told a little bit more, the reasons, to see how it works."

Overall, there are signs that ZCB could gain credibility with the farming community, but it is critical to maintain a respectful dialogue and acknowledge the importance of tradition and local expertise as a means to build those all-important bridges from science to practice.

About the author:

Sophie Wynne-Jones is a Research Associate with the Wales Rural Observatory where she does research to support Welsh Government with their rural and land use policy. Her webpage is http://www.aber.ac.uk/en/iges/staff/research/sxw/

5.4 ZCB and health

Seeking an environmental transition through health

Guppi Bola

This discussion piece calls for a renewed approach to a sustainable and just transition, one that recognises the value of a health enhancing and environmental protection principle known as Ecological Public Health.

Our aim should be to build an environment that puts population health at the heart of its actions towards sustainability, because the scale of the problems we face are considerable, and the impact of our inaction will be profound. Obesity has quadrupled in the last 25 years, inflicting over 22% of the adult population and set to increase to 50% by 2030 (Wang et al., 2011). It is also the most potent risk factor for type II diabetes, of which 5% of the UK population suffer (QOF, 2011). The economic burden of these illnesses combined has reached £5.8 billion a year and is rising, placing irrepressible strain on our healthcare system (Scarborough et al., 2011). But, more importantly, it has a considerable affect on individuals and families by increasing levels of depression, addiction and social isolation. We have failed to recognise that these so-called 'lifestyle' diseases are actually a product of the dimensions that we function in. Concordantly, emissions have risen by 4.5% over the past year (DECC, 2012), car dependency keeps families out of active transport and the population continues to rely on high carbon, high fat, highly processed meals. This is but a short demonstration of the interconnectedness of these issues, the difficulties of which will never be tackled without embedding public health principles into our future interventions.

Public health has been forgotten as one of the cornerstones of society's response to the environment and the way we function within it. From the early days of public health, environmental initiatives stemmed from the need to maintain the health of populations: achieving environmental protection and social justice as consequential outcomes. The work of John Snow, whose influence on water and sanitation procedures after the 1854 outbreak of cholera in London (Hempel, 2006) provides the best example of early public health measures. This was a landmark moment for epidemiology, where value was found in identifying the root cause of illness within groups of individuals, and not just in the individuals themselves. Snow was a pioneer; his investigation and recommendations enhanced the lives of those living around poverty-stricken Broad Street, as well as ensuring the protection of clean water from the tyranny of urban sprawl. Public health became a means of tackling societal concerns through health, and health concerns through the environment.

The discourse around health and the environment recognised that in many ways human activity was altering the natural landscape on which it survived. Over time these concerns eroded as they each became confined as issues requiring separate responses. Whilst health remained high in the public concern, medicine became the chosen route to tackle illness. The results of which created a resource heavy, industrialised health care system fixated with curative but not preventative practice. In addition, our attitude towards protecting the environment has come and gone, leaving politicians slow to respond, and our connection and value for nature diminished under unstoppable levels of urbanisation, motorised transport and unfettered consumption.

The mark of human activity has presented us with the challenges of climate change, resource shortage, biodiversity loss and deforestation. In order to respond effectively, we need to see the value of the health of our planet and the health of its people as the same thing. This is not to marginalise the term 'ecological' to one of a simple interaction between humans and the environment, but as Ernst Häckel suggested, have us accept the complex and multi-layered connections that this model presents (Krieger, 2001). John Hanlon, former Assistant Surgeon General of the US, said in the 1960s that public health needed to address the entire biological, material, social and cultural dimensions of the human, living, and physical world (IMNA, 2002). This was perhaps the first integrated presentation of ecological public health, one that embraced the complexity of interconnected dimensions.

Understanding these dimensions is the first step in designing an intelligent policy approach for public health and environmental sustainability. They are:

1) The material dimension – our physical and energy infrastructure (matter, energy, water). The biological dimension – the bio-physiological processes and elements (animal and plant species and also micro-organisms). The cultural dimension – how people think and what shapes their attitudes, spheres of interpersonal relationships, community, group and family traditions.
2) The social dimension – institutions created between people and expressed in terms of laws, social arrangements, conventions and the framework of daily living generally outside individual control (Lang and Rayner, 2012).

What we are missing is the ability to engage in all four dimensions of existence. A good example of this is our current approach to tackling the heavily processed food system that results in high levels of carbon emissions and diet-related illnesses. Telling families in low-mid income settings that they need to change to healthier diets ignores the social conditions that prevent them from doing so, and risks alienating those that are prime targets of food advertising, supermarkets and fast foods chains. A true ecological public health response that encourages meaningful behaviour change will recognise these interrelations through food education as well as through reshaping the food environment.

Achieving sustainable planetary, economic, societal and human health should be on the agenda of every activist, professional, politician and citizen across the UK. Given the current rhetoric around localism (bringing decisions back to a community level), this will work only if councils feel agency in pushing forward with radical approaches to sustainability. At the same time, any future action will be undermined by the political and economic determinants that are shaped at international and global levels. We need to break down these undemocratic structures in every dimension so that control is back in the hands of the public. But in order to succeed fully, professionals will together need to call for a mixture of interventions shaped by our environment that will ameliorate the determinants of health. If this works, we will have had the opportunity to rebuild our relationships with our surroundings and with each other, and refocus on the connections between our health and the environment for a just and sustainable transition for the future.

About the author:

Guppi Bola is a writer, researcher and activist working in the field of public health. She has spent the past six years bridging issues of the environment and health, working with academics, climate activists and medical professionals. Her time is now focussed on building the People's Health Movement UK to strengthen the fight against rising health inequalities. You can find her on Twitter @guppikb

5.5 ZCB and young people

Opportunities for young people in a zero carbon Britain

Louisa Casson of the UK Youth Climate Coalition (UKYCC)

Every young person has a stake in a future free from climate chaos. We've grown up with the knowledge that we need to make a change in the way our society works – and we have the creative thinking and the determination to make this happen.

The UK Youth Climate Coalition aims to inspire, empower, mobilise and unite young people to take positive action against climate change and become drivers for change towards a zero carbon future.

We see this happening by making sure there are opportunities for British youth to learn key knowledge and skills, and gain employment experience, to enable them to take a proactive role in a zero carbon economy.

The UKYCC envisions a world in which formal and non-formal education promotes an understanding of the issues of climate change and sustainability in young people, equipping them with the knowledge to apply this together in their daily lives. In this world:

The national curriculum prioritises a strong understanding of the scientific and ethical issues behind climate change.

An informed and educated understanding of climate change is vital for young people and future generations to live in the economy, society and environment of the next few decades. Climate change must be recognised as a political, cultural, economic and social issue and should be integrated across the curriculum. This will also fulfil the British government's commitment to Article Six of the United Nations Framework Convention on Climate Change regarding climate change education, and their duty to ensure that all children have universal access to a basic education of how climate change will impact on the world in which they will be living and working.

Everyone has the opportunity to participate in sustainable initiatives in their schools, colleges, universities, other non-formal learning contexts and wider communities.

Extra-curricular activities support young people to imagine and implement innovative sustainability solutions, enabling them to become change makers while still in formal education. In their wider communities, we envision a strong grassroots youth climate movement, which offers workshops and projects to channel the energy and creativity of young people into productive contributions to a zero carbon society.

Education prepares and empowers young people to work in the green economy, providing the skills needed to take up meaningful green jobs for all.

Both formal and non-formal education encourages young people to take active responsibility for their environment and equips them with knowledge and the capacity to make a valuable contribution to a zero carbon society.

The UKYCC envisions green jobs with fair pay and real progression opportunities made available to all young people as part of a strong green economy in the United Kingdom.

Green jobs give Britain the chance to offer meaningful employment opportunities to its citizens, who in return gain the opportunity to be at the heart of the transition to an environmentally sustainable economy.

We want to see young people's creative drive help shape the broadening of the definition of employment in a zero carbon economy. We define green jobs in four main ways:

- Green jobs are long-term, not just for the short-run. This can include training programmes and apprenticeships designed to lead to stable and long-term employment.
- Green jobs provide a way to improve your personal set of working skills and ensure future opportunities for building your career.
- Green jobs must be non-exploitative and provide a 'living wage', particularly with respect to young people in full-time employment for the first time.
- Green jobs are sustainable and have stewardship of the environment at their core.

New green jobs are created in all sectors, supported by investment, innovation and legislation.

Green jobs are not limited to traditional sectors such as waste management and renewable energy construction. We envision governmental support to build confidence across an ethical and environmentally sustainable economy, exploring the full range of possibilities for new forms of green employment in a zero carbon future. While 'direct' green jobs involve the manufacture, installation and maintenance of environmental technology, such as solar panels, we also envision the expansion of 'indirect' green jobs that are linked to maintaining and improving the quality of our natural environment – such as 'greening' existing jobs.

Everyone is supported to be a part of the transition to a zero carbon economy, including provision for individuals to make their current jobs and workplaces more sustainable.

Access to training and education is fair and open to people of all backgrounds and ages, ensuring that the opportunities of the green economy are open to everyone.

Education on sustainability and reducing carbon emissions is extended beyond young people to enable existing employment sectors to make effective changes to create green jobs across society. This involves reconceptualising the workplace, moving away from wasteful, carbon intensive workspaces by introducing or extending recycling facilities, low carbon travel initiatives, by limiting paper usage, using electricity more efficiently and encouraging sustainable lifestyle practices.

We believe that empowered youth must be at the centre of our pathway to a zero carbon future. Making green jobs and education opportunities readily available to young people will form the key building blocks for a healthy, just future.

About the author:

Louisa Casson works as a Communications Officer for the UK Youth Climate Coalition. Run entirely by a team of highly motivated volunteers aged 18-29, the UKYCC campaigns on issues closely relating to young people as part of a wider, cross-generational movement for change. The UKYCC believes that to tackle climate change we need an inspiring vision of how we want the world to be and a movement that anyone can feel part of.

5.6 ZCB and happiness

Why reducing consumption will make Britain happier

Liz and Mike Zeidler of Happy City Initiative

With planetary survival and economic meltdowns hitting the headlines, talk of increasing 'happiness' might sound like a pleasant and rather fluffy distraction. Yet growing numbers of people, across all sectors of society, are talking about happiness as if it really matters. Far from being a distraction, a focus on happiness may be the key to sustainable prosperity for all.

It seems to surprise people that Britain has seen 60 years of nearly constant economic growth, and yet during that time happiness and wellbeing measures hardly improved at all. This is not a coincidence.

Constant growth (as opposed to steady prosperity) requires each of us to be constantly a little bit unhappy – dissatisfied with what we have and thinking we always need more to be happy. Happiness must always appear to be just over the horizon, and more 'stuff' must be the pathway to it, in order for economic growth to continue year after year.

Whilst there is no correlation between perpetual economic growth and increasing happiness, there is a strong correlation between happiness and improving health, increased education, environmental improvements and reductions in inequality and crime. Happier people and populations consume fewer natural resources, learn better, and work more productively, creatively and collaboratively. They are healthier, less likely to be violent or act criminally, less dependent on welfare... the list goes on. In fact, it's a two-way correlation: improving society and the environment makes people happy, and making people happy improves society and the environment.

This isn't fluffy stuff. Happiness isn't just a pleasant outcome, or even 'just' an ultimate goal for society. As a way to reframe the debate and focus on what truly matters, happiness is also essential to solving most of our biggest challenges – including climate change.

'Demand reduction' is a major component of the Zero Carbon Britain scenario, and a major discussion point in Green circles. How do we get people to consume less stuff, be it energy, food, fuel or the latest electronic gadget? Demand reduction is an even more sustainable approach than simply pushing for more recycling and greener power. Lower demand means fewer resources are consumed in the first place, plus fewer emissions are released into the atmosphere in the process.

Demand reduction, however, has one big hurdle to overcome – it seemingly requires people to opt for a degree of self-sacrifice. I, the consumer, will 'give up what I'd like to have, for the greater good', – or, even more tenuously, 'for future generations'. It's a hard sell, particularly when balanced against the persuasive power of the multi-billion pound advertising industry urging ever-greater consumption. But the 'ultimate' demand reduction tool is surely to come at it from exactly the opposite angle to the one initially perceived – reducing demand not through self-sacrifice but through greater satisfaction with what we have.

And so a happiness revolution might be the answer to the negative perception. What could we achieve if we turned away from things that advertisers say will make us happy (and don't), and towards things that evidence shows will make us happy (and do)? It's certainly a much easier 'sell' to suggest we make ourselves happier through actions that just happen to reduce demand, rather than begrudgingly put up with less happiness for the greater good. When we reject the assumption that the route to happiness is via increased wealth and greater consumption and focus instead on the *real* pathways to happiness, we necessarily, naturally and happily consume less.

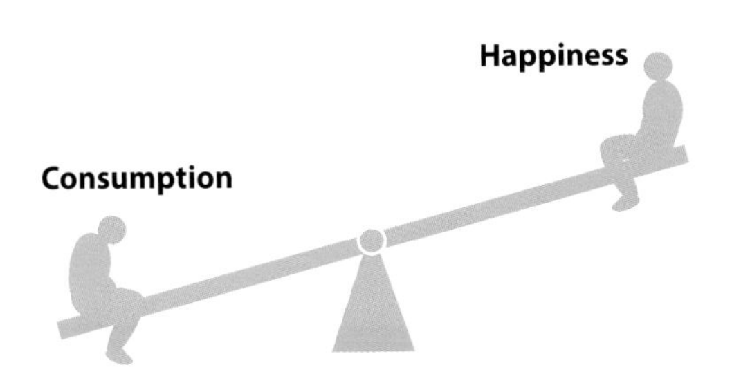

So what makes us happy? Ask someone what makes them happy, and whether they live in inner city Glasgow, rural Botswana, Palestine, Poland or Bristol, their responses will be remarkably similar. You'll hear talk of family, friends and community. A sense of belonging, purpose and value will all be high on the list. Also important are getting outside and living in a clean environment, as well as opportunities to interact and get involved, being active, learning and growing, helping others and being helped. None of these things need much 'stuff', and they show that a long-term, happy lifestyle is both low carbon and relatively low cost.

Paul Hawken estimates that "there are around 2 million organisations working toward ecological sustainability and social justice" worldwide (Hawken, 2007). In the UK alone, many thousands of exceptional and worthwhile organisations, projects and individuals are trying to tackle these big issues. But until we overcome the cause of our overconsumption – and acknowledge that our fixation on stuff is really an addiction to unhappiness – those activities can only ever be a sticking plaster on a deeper wound. Using the 'stick' of fear of the human or environmental consequences of our consumption isn't working on any significant scale. Selling the benefit from the 'carrot' of greater happiness is far more likely to persuade whole societies to change their behaviour and make different choices. Wouldn't you be more likely to forgo buying a new telly, or a new car, or a holiday abroad, if you knew you'd be happier for it?

We've been hooked on unhappiness for over 60 years, and it's going to take quite a bit of effort to quit. Everywhere we look, every magazine and billboard, every doom and gloom headline, every point scoring political jibe, is telling us we'll be happy when we have a new car, new hairstyle, new government.... We are riddled with fear, blame and a sense of lack.

So let us change the narrative and change the question. If prosperity means to flourish and to thrive, then people of all economic levels can demonstrate prosperity (and, for that matter, people of all economic levels can demonstrate poverty). We invite the people of Britain to redefine prosperity, to see it as a matter of contentment rather than consumption.

About the organisation:

Happy City Initiative exists to demonstrate that being happier needn't cost the earth. They enable a city-scale 'tipping point' of change through innovative community projects, campaigns and training, alongside new ways to measure prosperity in society. All this combines to create a city-wide new model of progress based on flourishing lives on a flourishing planet.

5.7 ZCB and Zero Carbon Egypt

ZCB and Zero Carbon Egypt

Isabel Bottoms

The idea to introduce a decarbonised development model for a developing country did not come from a general desire to spread 'zero carbonism', though that is a noble cause in its own right. It came instead from being inspired by the struggles of the young people in a specific country fervently working for a newly democratic, representative system. I wondered, how could I aid the transition? What skills or knowledge could I share to encourage economically and environmentally sustainable development for all Egyptians? What could be done to contribute to transitional justice? Through my interest in Zero Carbon Britain and in energy, food and poverty distribution in Egypt, the connection was made and a decarbonised development scenario was born.

The 'environmental' why?

There is nothing new in documenting the impacts of climate change on Egypt. As a large country in the Middle East and North Africa (MENA) region it will suffer (if it isn't already) from rising sea levels affecting populous port cities such as Alexandria and Port Said; salt water intrusion killing off crops, polluting the fresh groundwater sources and negatively impacting on freshwater fish species; droughts and flooding down the Nile affecting the majority of currently productive agricultural land in Egypt; increased waterborne diseases hitting the poorest in the population the hardest; increased desertification; plus rising temperatures and the potential for more extreme weather events, such as dust storms. The effects of many such environmental stresses on Egypt are those prominent in public discourse today: rising food prices, power shortages, rising fuel prices and loss of rural livelihoods.

Life on the ground

Egypt is an interesting case for the MENA region: once the breadbasket, it has been dubbed a "basket case" in recent media coverage. Egypt is now the second largest wheat importer in the world, but it used to feed itself. Egypt's cotton industry, famed for the finest cotton sheets, is in decline. Its people are also heavily reliant on natural gas for their heating and cooking, which is very cheap as long as the subsidies remain in place. Diesel is becoming increasingly scarce, and farmers, hauliers and other key actors in the food chain have to spend up to a day queueing for their share of the subsidised, and soon to be rationed, diesel.

Since President Sadat's open door policy of the eighties, businesses big and small have gone about their profit-making ventures with little regulation. This means international companies and development projects, as well as state-owned and local enterprises, are damaging local livelihoods and environments by polluting the groundwater (used for drinking and irrigation), destroying agricultural land by building on it, and by contributing to widespread pollution of the Nile. Meanwhile, current government fuel and bread subsidies cannot be sustained and Egypt's economy is up to its eyes in debt and riddled with inefficiencies.

The political why?

The uprisings in Egypt over the last two years are indicative of many factors – resource scarcity and poverty being two very important ones. These affect the everyday lives of millions across Egypt, whose anger at the politics dictating their lives can only be expected to increase as their situations deteriorate. Their demands cannot be ignored, yet neither can the science predicting how badly they will be affected if climate change is not mitigated. Current social structures must also become resilient to the worst and unavoidable effects of climate change.

The time is ripe for a set of co-ordinated policies with deep social and environmental integrity. Politically active Egyptians across the the political spectrum advocate, criticise and discuss what befits them and what serves their varying political ends, but they have yet to articulate objectives that are both socially just and environmentally sound. A unique opportunity exists to contribute to the transitional dialogue between Egyptians, in the hope of creating a legacy that will serve Egypt's current and future generations.

How are we doing it?

From the germ of the inspiration, there is now a 'we' making this happen. With support from the Zero Carbon Britain team, the first phase – the research – of the Egyptian project will be conducted by the Desert Development Centre within the American University in Cairo. Using the same techniques and a zero carbon framework, potential scenarios will be developed through a carbon model. The difference for the Egypt process is that, in consultation with Egyption stakeholders and experts, there also needs to be a defined 'development space'. Through representative stakeholder groups and expert advice this 'space', which defines Egypts developmental trajectory and potential, will be defined from the outset. This allows for the backcasting and scenario building that incorporate Egypt's aspirations and needs. Furthermore, every policy will come up against a set of criteria which represent social justice and developmental elements. Without jumping to conclusions, these are likely to be efficiency, effectiveness, and equity – again, terms which will have to be defined by Egyptian stakeholders from the outset.

The second phase is where the project aims to get real leverage. That is, translating technical and alienating environmental and scientific jargon into a framing that the average Egyptian person will relate to. Having identified social justice as the tagline of the opposition parties, the youth and the general population, it will be the method by which this zero carbon scenario is framed. Social justice can include job creation, energy security, food security and access to efficient and safe housing. To do this the project is partnering with the Egyptian Centre for Social and Economic Rights, as they have established working groups in most of Egypt's Governerates and can offer access to unions, farmers, workers, local communities, etc. They will also carry out the reframing and writing of the zero carbon scenario research within social justice terms.

The third phase is to catalyse widespread take up of the findings and solutions offered in the report by advocating on all levels of Egyptian society. Using strategic power mapping, systems intervention analyses, sector specific micro-models of how to implement the report's recommendations, and with convincing business models we hope to reach the huge informal sector in Egypt as well as the highest levels of government. We will also be promoting the cultural aspects surrounding the adoption of the recommendations.

Egypt is begging for opportunities and pathways to a better future, and we are now seeking funding to be able to offer it a sustainable and zero carbon future.

About the author:

Isabel Bottoms is a young Welsh woman inspired to initiate Zero Carbon Egypt by Egyptian friends active within, and affected by, the uprisings of the last two years. Drawing on her experience as a youth climate activist, graduate of law, equity advocate, and policy and strategy geek, she's co-ordinating the project.

5.8 ZCB and creative practice

The Great Imagining: making art as if the world mattered

Lucy Neal

It's believed the 'Lion Man' statuette found in a cave in the Swabian Alps (British Museum, 2013), took someone 400 hours to carve 40,000 years ago. With a human body and a lion's head, it is evidence of our early ability to conjure in our 'mind's eye', giving form to things we first imagine. Whether or not art then played a role in communicating with the supernatural, social ritual, song or storytelling, it shows an impulse to articulate our place and survival in the web of life.

Today, ZCB asks how our 21st century skills can help us rehearse the survival of our own and other species over the coming decades. It squares up to the challenge that will dominate the arts and culture, and all of society, for the foreseeable future: our human relationship with energy. The report gives clear evidence that business as usual is not an option. Decarbonisation scenarios show how rapid change in our homes, transport, food production and lifestyles is possible. It is a framework for life in our current unsustainable energy system (which we Power Down) that imagines and creates an alternative (which we Power Up). It feels an intense task to live two 'realities' at once like this – but, if we keep our focus, it can change world orders. It also has precedents: playwright Vaclav Havel called it 'living in truth' – living 'as if' he were a free man, even when imprisoned. He saw the alternative in his mind's eye and, by believing it, brought that reality into being.

"Believe you can and you're halfway there."
Theodore Roosevelt

ZCB's robust research gives the building blocks to live our lives 'as if' a zero carbon Britain by 2030 were possible: it gets us 'halfway there'. In the arts, it has already inspired new work – Kaleider in Devon invites artists to imagine a future 30 years from now:

"We use this idea of an imagined future to pull artists into a creative space where they might be able to deal with some of the big challenges without freezing in the face of their magnitude: we can ask open questions without a need to campaign."
Seth Honor, Kaleider

ZCB's clarity "enables us to grasp the sense of urgency with which we have to tackle carbon reduction" and "the role that all those working with arts and culture can play in achieving that critically important goal", says Clare Cooper of Mission Models Money. In turn, the arts and culture have a distinct role to play in inspiring a wholly different way of living within the ecological limits of a finite planet, and in remodelling society – reinventing its rules and values. The arts have a tradition of sparking cultural change and 'speaking differently' (Prof.

Richard Rorty). There's a sense that science and technology alone cannot play the role of interpreting the challenges we face or questioning what values underpin the need for change.

On the surface, contemporary mainstream culture appears unable to conjure with pictures of the future that are not apocalyptic. ZCB releases us psychologically from this grip to start the 'great imagining' – creating a zero carbon future. "It's a foundation stone to jump from, to imagine crossing the yawning chasm, between where we are now and where we need to be", says Teo Greenstreet, Case For Optimism. It "supports our aims and trajectory with huge amounts of data – facts and figures that are trustworthy and relevant", adds Feimatta Conteh from the Arcola, a progressive theatre aiming at being zero carbon, designing hydrogen fuel cell power systems alongside the staging of plays.

"... it's the job of the artist, poet or storyteller to point out the ground under our feet, to offer us images through which to wake up to our present condition, to show us anew the moment we stand in."
Mat Osmond, *Dark Mountain* Issue 3

Initiatives such as Tipping Point, Cape Farewell, Julie's Bicycle, Platform, Creative Carbon Scotland and the Centre for Alternative Technology's own Emergence Summit have set the pace in recent years, galvanising awareness of 'the moment we stand in'. Artists, in collaboration with scientists and energy specialists, have created a movement in the cultural sector that accepts responsibility for embracing the urgency of climate change whilst maintaining the poet's instinct to come at things sideways – as the playwright Chekhov said, "Don't tell me the moon is shining: show me the glint of light on broken glass".

For me, as a theatre-maker and writer, ZCB helps me to cross the line connecting my creativity to social and ecological responsibilities, and I'm excited about this. An original purpose for art to create value and meaning in our daily lives feels renewed and relevant. The 'engaged optimism' of movements such as Transition Towns shows that transformation of society becomes an art in itself, while Encounters shops on the high street – where 'Nothing is on sale, but lots on offer' – ask, 'What is it like to live now?' Participation across generations and cultures unearths deep emotional connections between people and helps us to express fears for the world, as well as hope and joy for the future. We extend a belief in ourselves as capable of remaking the world the way we would like it to be: creative, connected, happier and more resilient. Learning and 'qualities of neighbourliness' across difference become the foundations for what Barbara Heinzen (2004) recognises as rehearsal for social reinvention. A viable future can be created collectively but it must first be imagined.

ZCB invites – indeed, ignites – discussion about what kind of society we want to live in and the role the arts and culture have to play. A shared practice across science, art, politics, food growing, health and education can push for whole government policy frameworks for 2030 that connect cultural, social, environmental and economic frameworks 'as if the world mattered' (Suzi Gablik, 2002).

Like the craftsman or woman who sat carving the Lionman, hour after hour, we must keep our focus on creating the world we want from our collective imagination.

"Humans are capable of a unique trick, creating realities by first imagining them, by experiencing them in their minds. ... By this process it begins to come true. The act of imagining somehow makes it real. .... And what is possible in art becomes thinkable in life."
Brian Eno

About the author:

Lucy Neal OBE is author of *Playing For Time*, a handbook of transitional arts practice and acts of creative community to be published in 2014. Co-founder Director of the LIFT Festival (1981-2005), she is currently active in Transition Town Tooting.

End notes, glossary and references

Find out more

A wide variety of additional materials and resources linked to the Zero Carbon Britain project can be found on our website **www.zerocarbonbritain.org**

These include:

- ZCBlog posts containing up to date news on the ZCB project.
- Our methodology – how we constructed our scenario, including details on the data and references on which our research was based.
- More detailed background information on technologies and concepts related to our scenario.
- Answers to Frequently Asked Questions (FAQs) relating to our scenario and the ZCB project.
- Downloadable resources for teachers, policymakers, civil society organisations, campaign groups and other interested parties.
- Further '*ZCB and...*' discussion papers written by a variety of individuals and organisations.
- Our contact details and ways for you to get involved in the project.

We also include links to related projects or organisations which support, provide further information on, or work towards a zero carbon future.

The Centre for Alternative Technology offers practical and academic courses on various topics related to zero carbon and sustainable living through its Graduate School of the Environment (GSE), hosted at the Wales Institute for Sustainable Education (WISE) – an award winning venue with outstanding sustainability credentials. CAT also provides educational tours, workshops, outreach activities, day visits and residential visits (staying in CAT's Eco-cabins) for school groups, universities and educators. We offer consultancy and free information services relating to renewable energy and energy saving techniques, and have published a wide range of books on sustainable living.

The Home Energy Handbook, published by CAT in 2012, offers practical advice on how to save and generate energy in homes and communities in the most carbon- and cost-effective ways – an ideal guide for those seeking smaller-scale solutions to the climate and energy challenges addressed by ZCB.

To receive our quarterly magazine *Clean Slate* why not become a 'CAT member' and help support our work to **inspire, inform and enable**. Find out more about any of the above at **www.cat.org.uk**

Notes

Units

Here is a list of common units we use in this report and what they mean.

°C	degrees Celsius; temperature measurement.
g	gram; unit of weight.
ha	hectare; unit of area of land.
kcal	kilocalorie; energy contained in food.
m	metre; unit of distance.
mph	miles per hour; speed – how fast something is travelling.
MW	megawatt; unit of power; the rate at which energy is produced or used.
MWh	megawatt-hour; unit of energy.
tCO_2e	tonne of carbon dioxide equivalent; a measure of greenhouse gas (GHG) impact relative to carbon dioxide (CO_2). For example:

- Carbon dioxide (CO_2) = 1 x CO_2
- Methane (CH_4) = 21 x CO_2
- Nitrous oxide (N_2O) = 310 x CO_2
- Super GHGs = 150 to 23,900 x CO_2

Natural units

Ever wondered what we mean by a TWh? Yes, it's a million megawatt-hours, but what does that actually *mean*? Here are some (hopefully!) helpful examples.

Greenhouse gases (GHGs)

tCO_2e – one tonne of CO_2e. About 1.2 tCO_2e of GHGs are emitted during a return flight from London to New York; or, about 2 tCO_2e are emitted in the annual commute of one person travelling alone by car from the outskirts of London to the city centre.

$ktCO_2e$ – one thousand tCO_2e. Almost one $ktCO_2e$ would be emitted if we flew the entire UK Olympic squad (541 athletes) around the world once.

$MtCO_2e$ – one million tCO_2e. The city of Oxford was responsible for just under 1 $MtCO_2e$ of emissions in 2003. It is estimated that about 13.1 $MtCO_2e$ is emitted during a year's worth of commuting in the UK.

$GtCO_2e$ – one thousand million tCO_2e. In 2005, global GHG emissions totalled about 45 $GtCO_2e$.

Energy

MWh – one megawatt-hour. A typical UK household consumes around 4 MWh of electricity per year.

GWh – one thousand megawatt-hours. The total energy consumption of Cornwall in 2007 was 12,026 GWh; one supermarket uses about 2.5 GWh of electricity per year.

TWh – one million megawatt-hours. The UK's daily electricity consumption is a bit less than 1 TWh.

Land area

ha – 100 metres by 100 metres. Trafalgar Square in London is about 1.2 ha; a football pitch is about 0.7 ha.

kha – one thousand hectares. The area of Manchester is about 11.5 kha while that of Norwich is only about 3.9 kha.

Mha – one million hectares. The area of Belgium is about 3 Mha. The UK's area, including coasts, rivers and lakes, is about 24.7 Mha.

Acronyms

AD	Anaerobic digestion
BECCS	Bio-energy carbon capture and storage
CCS	Carbon capture and storage
CHP	Combined heat and power
DECC	Department for Energy and Climate Change
FAO	Food and Agriculture Organization
GDP	Gross domestic product
GHG	Greenhouse gas
GM	Genetically modified
HFSS	High fat, salt and sugar
HGV	Heavy goods vehicle
HPI	Happy Planet Index
IPCC	Intergovernmental Panel on Climate Change
NDNS	National Diet and Nutrition Survey
NI	Nitrogen inhibitors
NPS	Nutritional profile scores
PCA	Personal Carbon Allowance
PV	Photovoltaic
REA	Renewable Energy Association
RUK	RenewableUK
TEQs	Tradable energy quotas
UK	United Kingdom
UKCIP	UK Climate Impacts Programme
UNFCCC	United Nations Framework Convention on Climate Change

Glossary

Adaptation – changes that we make to natural or human systems (infrastructure, political systems) to minimise, adjust to, or take advantage of the effects of climate change.
Ambient energy – low temperature heat energy in the air, the ground and water. Can be extracted and 'concentrated' to higher temperatures by *heat pumps*.
Anaerobic digestion (AD) – breakdown of plant material, food wastes and manure by bacteria which produces *biogas* that contains *methane* (CH_4).
Atmosphere – layer of gases around the Earth that protects us by absorbing solar radiation, warms by keeping heat in (via the *greenhouse effect)*, and reduces temperature extremes between day and night.

Back up generation – form of electricity generation used when not enough energy is available, usually a form of *dispatchable generation*.
Biochar – virtually pure carbon derived from *biomass* through the process of *pyrolysis*. A portion of the carbon remains stable (not *biodegradable*) for hundreds to thousands of years.
Biodiversity – from biological-diversity; variety in the natural world, including variations within and between species, *ecosystems* and *habitats*.
Biodegradable – compostable, material that decomposes or breaks down back to basic elements.
Biofuel – liquid fuel made from *biomass*.
Biofuel, first generation: biofuel produced from crops such as wheat, corn, sugar crops or vegetable oil.
Biofuel, second generation: biofuel produced from woody material, such as fast-growing trees and grasses.
Biogas – gas containing *methane* (CH_4), the carbon in which originates from recently grown *biomass*. Biogas can also contain impurities such as CO_2, which when removed leave pure or near pure methane. The *methane* in biogas produces energy when burned (like fossil fuel gas).
Biomass – plant and animal material.
Bioreactor – manufactured, engineered or controlled environment designed to encourage decomposition of plant material, usually by adding air or liquid. The gases produced can be captured and used to produce energy.

Cap and Share – *downstream emissions reduction scheme* where a '*hard cap*' is placed on emissions produced by energy suppliers. Emissions permits are shared equally per capita among the adult population.
Carbon budget – or **cumulative carbon budget**; an amount of carbon dioxide (CO_2) or *greenhouse gas* that can be emitted over a budget period. Carbon budgets are used to define the maximum emissions that can occur before there will be a particular risk of various degrees of *climate change*.
Carbon capture – the taking in of carbon (usually CO_2) by natural systems, usually (though not always) through *photosynthesis*. (In this report, the opposite to carbon *emission*).
Carbon capture and storage (CSS) – process of capturing CO_2 emitted as waste (from fossil fuel power plants, for example) and storing it, normally underground or underwater (in old oil or gas fields, for example) to prevent it being released into the *atmosphere*.
Carbon cycle – movement of carbon through the land, oceans and atmosphere in various different forms (for example, CO_2, or carbon in plants).
Carbon dioxide (CO_2) – the primary *greenhouse gas* emitted by human activities. It is the largest contributor to climate change.
Carbon flow – movement of carbon around the *carbon cycle*, for example, *carbon capture* of *carbon dioxide* by plants during *photosynthesis*.
Carbon intensity – amount of carbon emitted to produce a unit of output.
Carbon neutral – GHG emissions are balanced by *carbon capture* such that the net emissions are zero, or neutral.
Carbon neutral synthetic liquid fuel – man-made fuel from the combination of hydrogen and carbon using the *Fischer Tropsch process*. Hydrogen is obtained by *electrolysis* using electricity from a

renewable source, and the carbon comes from *biomass*, making the *fuel carbon neutral*.

Carbon neutral synthetic gas – man-made fuel from the combination of hydrogen and carbon using the *Sabatier process*, where the hydrogen is obtained by *electrolysis* using electricity from a renewable source, and the carbon comes from *biomass*, making the gas carbon neutral.

Carbon store – a place where carbon can be kept out of the *atmosphere* for a significant period of time (for example, carbon in the plant matter of trees, or in soils).

Carbon tax – an *emissions reduction scheme* where a tax is paid on activities that cause *greenhouse gas* emissions.

Climate – defines what the 'normal' and 'extremes' of weather are in a region. Climate is usually defined as 'an average of weather' over about 30 years. Though different places have different climates, the globe as a whole has a defined climate, averaged over all locations.

Climate change – change in global *climate* as a result of increased levels of *greenhouse gases* in the atmosphere (largely from burning *fossil fuels*) that enhance the *greenhouse effect*, causing warming and other impacts.

Combined heat and power (CHP) – systems in which the combustion of fuels generates usable electricity and also heat. Common in industry and for community heating schemes.

Compost – decomposed organic (plant derived) material used as *fertiliser* for soil.

Consumption emissions – *greenhouse gas* emissions from the production of all goods and services consumed by a nation. Includes *greenhouse gas* emissions from goods and services produced for, but not within, a nation (imports). Excludes the *greenhouse gas* emissions from the production of goods and services that are exported.

Contrails – long thin artificial clouds that sometimes form behind aircraft.

Cumulative carbon emissions – sum of *greenhouse gases* emitted year-on-year, creating a total that represents all GHG emissions over a period of time.

Decarbonise – to remove the GHG emissions from a product, service or system by changing the way it is produced or operates.

Demand management – shifting energy demand from times when energy supply is low to times when energy supply is in excess.

Denitrification – the release of nitrous oxide produced when microbes act on nitrogen deposited in the soil by fertilisers.

Dispatchable generation – a form of electricity generation that can be called upon to operate as and when required, for example, as *back up generation*. Ideally, power stations that provide dispatchable generation can increase or decrease output quickly and without efficiency losses.

Downstream – a system whereby the focus is on individuals to change their behaviours (driving, flying, etc.) to reduce GHG emissions.

Ecosystem – a system formed by the interaction of a community of organisms (plants, animals, etc.) and their environment (for example, a chemical system like the water or *carbon cycle*).

Electricity grid ('the grid') – a system of wires and equipment that transports electricity from generators to consumers. The grid must be 'balanced' so that electricity supply matches demand.

Electrolysis – the process of 'splitting' water (H_2O) into hydrogen (H) and oxygen (O) using electricity.

Emissions allowance – emissions permitted by an individual, organisation or nation as designated by an international agreement or *emissions reduction scheme*.

Emissions pledge – amount by which a nation has promised (sometimes set in law) to reduce its emissions relative to a particular year, usually by a certain date (an emissions reduction target).

Emissions reduction scheme – a policy framework designed to reduce *greenhouse gas* emissions.

Emissions cap – total permitted *greenhouse gas* emissions as set by international agreement, government or organisation, usually on an annual basis, resulting in year-on-year reductions.

Emissions trading scheme – a *'soft cap' upstream emissions reduction scheme* where permits to emit

greenhouse gases are distributed to emitters – mainly industry and businesses. Permits can be traded.
Energy intensity – amount of energy required to produce one unit of output.
Energy crop – crop grown and harvested specifically for the production of energy.
Energy demand – or **final energy demand**; the amount of energy required/consumed, excluding conversion and distribution losses. In this report, this is the same as *final energy demand*.
Energy supply – or **primary energy supply**; the 'raw' energy input before any losses from conversion or transmission processes.
Energy use – refers to energy used by a final user. This excludes conversion and distribution losses, but includes end use inefficiency losses (for example, energy lost as heat by electrical appliances).
Enteric fermentation – occurs during the digestion of food by a cow or sheep (or other ruminant). Methane is one of the by-products of this process.

Fertiliser – provides the necessary nutrients required for plant growth (in addition to sunlight and rain) when applied to the soil. The most common nutrients are *nitrogen*, potassium and *phosphorous*.
Fossil fuel – material made over the course of hundreds of millions of years from plant and animal material that has been heated and compressed by various natural geological processes. The burning of fossil fuels emits additional *carbon dioxide* into the *atmosphere* and contributes to *climate change*.
Fischer-Tropsch process – a chemical process that uses carbon monoxide (CO) and hydrogen (H) to form synthetic liquid fuels.
Fixed offshore wind turbine – offshore wind turbines with foundations embedded in the seabed, in contrast to *floating offshore wind turbines*.
Floating offshore wind turbine – offshore wind turbines floating in the water and connected to the seabed using anchor cables. Can be used in deeper water than *fixed offshore wind turbines*.
Fracking – or 'hydraulic fracturing': the unconventional extraction of oil which involves inserting a mix of chemicals under high pressure into an area underground to release the fossil fuel gas trapped in shale.
Fuel mix – the types and quantity of fuel required by *energy demand*.

Global average temperature – the average temperature of the Earth's surface, as measured combining thousands of temperature measurements on land and on sea.
Greenhouse effect – the warming of the Earth's surface due to the absorption and reflection of heat leaving the Earth by *greenhouse gases* in the *atmosphere*.
Greenhouse gas – a gas in the *atmosphere* that absorbs heat from the Earth and emits it in all directions.

Habitat – a particular area or environment inhabited by a species, plant or animal.
'Hard cap' – emissions are not allowed to exceed an agreed/designated limit (the 'cap').
Heat pump – a technology that extracts and 'concentrates' *ambient heat* from a low temperature source (the air, water or the ground) and delivers it as useful heat at a higher temperature.
Heat recovery ventilation – a type of ventilation in which the heat from exhaust air is transferred to incoming fresh air without the two air sources combining. This reduces both heat lost by ventilation and space heating demand.
Heat store – electricity is used to warm a tank of water, for example (the'heat store'), so that heat is available for later use.
Heat stress – the detrimental impact felt by plants and animals (including humans) when temperatures are too high, or they remain high for long periods of time.
Historical responsibility – the responsibility taken on for GHG emissions in the past when calculating *cumulative carbon emissions* measured against a nation's *carbon budget*.
Hydropower – generating electricity from water flowing downhill.

Industrial emissions – emissions of *greenhouse gases* that are produced by industrial processes (but not

related to energy production), usually as a result of chemical processes.
Industrial output – the amount of products produced by industry. It can be measured in monetary value by weight or volume – tonnes of steel, for example.
Infrastructure – physical and social structures that make our society work (for example, roads and electricity grid, or governmental systems)
Insulation – material used in the fabric of buildings to reduce heat loss.
Intensively grazed grassland – grassland that is managed intensively to graze livestock (usually sheep and cows), which is often fertilised.

Kyoto Protocol – international agreement to reduce *greenhouse gas* emissions under the UNFCCC.

Livestock – animals kept to produce meat or dairy products (usually cows, sheep, pigs and chickens).

Methane – flammable gas with the chemical formula CH_4. It is the chief component of the fossil fuel 'natural gas' but is also produced from biological material in *anaerobic digestion* and other processes (see *biomethane*).
Miscanthus – also known as 'elephant grass', a tall grass harvested usually every year as an *energy crop* with a high *yield*. Used as biomass for producing *biogas*, *biofuel* or *synthetic fuels*.
Mitigation (of climate change) – actions to limit the impact, or rate of, long-term climate change; usually involves the reduction of *greenhouse gas* emissions.
Monoculture – single plant species (an area that is planted with a monoculture is low in *biodiversity*).

Net energy importer – where more energy is imported than exported.
Nitrogen – a chemical element needed for plant and animal growth. Found in *fertilisers*.
Nitrogen inhibitors – chemicals that block the conversion of nitrogen to nitrous oxide in soils, thereby reducing nitrous oxide emissions.
Nitrous oxide – a *greenhouse gas* with a greenhouse effect roughly 21 times that of *carbon dioxide*.
Non-CO_2 emissions – *greenhouse gas* emissions that are not in the form of carbon dioxide (CO_2). For example, *methane* (CH_4), *nitrous oxide* (N_2O) and *super greenhouse gases*.
Nutrients – substances that provide essential components required for life. These can be minerals for plants, or vitamins required for humans.

Ocean acidification – the process of ocean water becoming more acidic (usually through CO_2).
Offshore wind – electricity production from either *fixed* or *floating offshore wind turbines* situated out at sea.
Onshore wind – electricity production from wind turbines on land.

Passivhaus – a building certified as complying with the Passivhaus standard requires buildings to have a very low heating demand (15 kWh per metre square of floor area per year, or less).
Peak oil – the point at which maximum extraction of oil is reached, and conventional supply sources go into decline.
Peat – type of soil that contains a high level of dead organic matter (plant material) that has accumulated over thousands of years.
Peatland – area of land where *peat* is found.
Permafrost – soil at or below freezing point (0°C) for two or more years.
Personal carbon allowances – a *downstream emissions reduction scheme* where *emissions allowances* are allocated equally per capita within a given population.
Phosphorus – chemical element that is essential for life; low levels can limit growth.
Photosynthesis – the conversion of sunlight into energy by plants. A plant takes in carbon dioxide and uses the carbon to grow new plant material.
Power to gas technology – technology that uses electricity to produce gas. For example, (surplus) renewable electricity can be used to produce hydrogen and, in a subsequent step using the *Sabatier reaction*, methane gas.
Pre-industrial – usually cited as before c. 1750 when

the industrial revolution began.
Production emissions – includes *greenhouse gas* emissions from all activities occurring in a territory but excludes emissions from goods and services produced outside the territory but which are consumed within the territory (imports).
Projection (of climate change) – indication from climate modelling of what is likely to happen in the future with respect to global (or regional) climate.
Pyrolysis – the heating of biomass at high temperatures in the absence of air to produce *biochar* and *biogas*.

Renewables – technologies that use renewable sources of energy – that is, those which are continually replenished, such as sunlight, wind, rain, tides, waves and geothermal heat.
Retrofitting – the improvement of existing buildings with energy efficiency measures, such as insulation, better windows and doors, draughtproofing and *heat recovery ventilation*.

Sabatier process – a chemical process that uses hydrogen (H) and carbon dioxide (CO_2) to produce methane gas (CH_4) and water (H_2O).
Semi-natural grassland – grassland that is managed to some extent, though not intensively. Covers a wide variety of *habitats* and is a good *carbon store*. Currently, a large proportion of semi-natural grassland is grazed by *livestock*.
Short Rotation Coppice (SRC) – usually made up of willow and poplar species which are 'coppiced' (cut back) after a few years. Coppiced *biomass* can be used to produce heat, for producing *biogas*, *biofuel* or *synthetic fuel*.
Short Rotation Forestry (SRF) – usually made up of fast-growing species of trees, such as birch, alder and sycamore which are planted and harvested regularly, usually for use as *biomass* for heat.
Smart appliances – electrical appliances with controls that allow them to alter the pattern of operation and thereby assist the balance of the *electricity grid*.
'Soft cap'– emissions above an agreed/designated limit (the 'cap') are allowed, but prices discourage behaviours that may cause this to happen.
Soil carbon – carbon stored in soils. Can be taken in by soils directly, or transferred through the carbon in litter from plants (dead leaves, branches, etc.).
Solar photovoltaic (PV) – technology producing electricity from the energy in sunlight.
Solar thermal – technology producing heat from the energy in sunlight.
Storage silo – landfill sites can be converted into storage silos so that decomposition of materials is almost entirely stopped, thereby preventing the emission of *greenhouse gases*.
Sustainability – the potential for long-term maintenance of *wellbeing*, dependent on the surrounding environment, economics, politics and culture.
Sustainably managed woodland/forest – woodland or forest that is harvested for timber to produce wood products and is replanted after felling, maintaining *biodiversity*.
Super greenhouse gas – *greenhouse gas* that has a much stronger warming effect than CO_2.
Synthetic fuel/gas – man-made fuel from the combination of hydrogen and carbon: in contrast to fuels with a fossil base (for example, petrol) or *biomass* base (for example, oil seed crops).

Temporary grassland – grassland that is usually harvested on an annual basis; can form part of a crop rotation.
Tradable Energy Quotas (TEQs) – *downstream 'hard cap'* scheme for limiting *greenhouse gas* emissions. Government sets the cap and a proportion of emissions are allocated to adult household members. The rest of the emissions permits are sold to non-household energy users.
Tidal stream energy – energy created from marine currents caused by changing tides, typically harnessed using underwater turbines.
Tidal range energy – energy created from the difference between high and low tides, typically harnessed by turbines in the walls of structures (barrages or artificial lagoons, for example) that hold back tidal water.

Unconventional oil – oil accessed by unconventional means (for example, *'fracking'*), as opposed to from an oilfield or oil well.
Upstream system – a system whereby the focus is on energy suppliers and fossil fuel users to decrease GHG emissions.

Waste emissions – emissions that are a by-product of a process or system.
Weather – short-term (day-to-day) changes in temperature, rainfall and humidity.
Weather systems – atmospheric dynamics (like pressure and temperature) that typically bring certain types of *weather.*
Wellbeing – social, economic, psychological, spiritual or medical welfare of an individual or group.
Wildlife corridor – an area of habitat (woodland, grassland, etc.) connecting wildlife populations that have been separated by human developments (roads, trainlines, etc.).

Yield – output (for example, energy, biomass or food crop) produced per unit of land.

References

Chapter 2. Context

2.1.2 Climate change

Allison, I. et al. (2009) The Copenhagen Diagnosis 2009: Updating the world on the Latest Climate Science, November. The University of New South Wales Climate Change Research Centre (CCRC). http://www.copenhagendiagnosis.com

Anderson, K. (2012) 'Real Clothes for the Emperor: facing the challenges of climate change'. Tyndall Centre, University of Manchester.

Anderson, K. and Bows, A. (2010) 'Beyond 'dangerous' climate change: emission scenarios for a new world', 29 November 2010, *Philosophical Transactions of the Royal Academy of Science* 369(1934), pp. 20-44, 13 January. DOI:10.1098/rsta.2010.0290.http://rsta.royalsocietypublishing.org/content/369/1934/20

Ballantyne, A. P. et al. (2012) 'Increase in observed net carbon dioxide uptake by land and oceans during the past 50 years', *Nature*, 488(7409) pp. 70-2. http://dx.doi.org/10.1038/nature11299

Baumert, K. A. et al. (2005) Navigating the Numbers: Greenhouse Gas Data and International Climate Policy. World Resources Institute.

Church J. A. and White N. J. (2011) Sea level rise from the late 19th to the early 21st Century. *Surveys in Geophys* (2011). doi:10.1007/s10712-011-9119-1

Coumou, D. and Rahmstorf, S. (2012) A decade of weather extremes, *Nature Climate Change*, 2, pp. 491-496.

Dai, A. (2012). Increasing drought under global warming in observations and models, *Nature Climate Change*, 3(171), pp. 52-58. DOI:10.1038/nclimate1633.

Hansen, J. et al. (2011) Climate Variability and Climate Change: The New Climate Dice, 10 November. NASA. [PDF] http://www.columbia.edu/~jeh1/mailings/2011/20111110_NewClimateDice.pdf

Hansen, J. et al. (2010) Global surface temperature change, *Rev. Geophys.*, 48, RG4004, doi:10.1029/2010RG000345.

IPCC (2012) 'Summary for Policymakers'. In: Managing the Risks of Extreme Events and Disasters to Advance Climate Change Adaptation [Field, C.B., V. Barros, T.F. Stocker, D. Qin, D. J. Dokken, K. L. Ebi, M.D. Mastrandrea, K.J. Mach, G-K. Plattner, S. K. Allen, M. Tignor, and P. M. Midgley (eds.)], A Special Report of Working Groups I and II of the Intergovernmental Panel on Climate Change, pp. 1-19. Cambridge University Press, Cambridge, UK, and New York, NY, USA.

IPCC (2007) Climate Change 2007: Synthesis Report, Contribution of Working Groups I, II and III to the Fourth Assessment Report of the Intergovernmental Panel on Climate Change [Core Writing Team, Pachauri, R.K and Reisinger, A. (eds.)]. IPCC, Geneva, Switzerland.

Lemos, M. C. and Clausen, T. J. (2009) Synthesis Report from Climate Change: Global Risks, Challenges & Decisions (Second Edition), 10-12 March. University of Copenhagen, Copenhagen. [PDF] http://climatecongress.ku.dk/pdf/synthesisreport

Lewis, S. L. (2011) 'The 2010 Amazon Drought', *Science*, 331(6017) pp. 554, 4 February 2011. DOI: 10.1126/science.1200807. [Abstract] http://www.sciencemag.org/content/331/6017/554.abstract?sid=2b48efd6-3837-45a4-8210-94cc9c7498f7

McCandless, D. (2010) 'Information is Beautiful: When Sea Levels Attack', the Guardian Data Blog, 22 February. http://www.guardian.co.uk/news/datablog/2010/feb/22/information-beautiful-sea-level-rise-climate-change#

McMullen, C. P. (Ed.) (2009) 'Chapter 4: Earth's Ecosystems', in: Climate Change Science Compendium 2009. United Nations Environment Programme. [PDF] http://www.unep.org/pdf/ccScienceCompendium2009/cc_ScienceCompendium2009_ch4_en.pdf

Meehl, G. A. et al. (2007) 'Chapter 10: Global Climate Projections', in: Climate Change 2007: The Physical Science Basis. Contribution of Working Group I to the Fourth Assessment Report of the Intergovernmental Panel on Climate Change [Solomon, S., D. Qin, M. Manning, Z. Chen, M. Marquis, K. B. Averyt, M. Tignor and H. L. Miller (eds.)]. Cambridge University Press, Cambridge, UK and New York, NY, USA. p. 783. 'Frequently Asked Question 10.1 Are Extreme Events, Like Heat Waves, Droughts or Floods, Expected to Change as the Earth's Climate Changes?' http://www.ipcc.ch/publications_and_data/ar4/wg1/en/faq-10-1.html

Met Office (2011a) The impact of a global temperature rise of 4°C. Met Office, 29 June, http://www.metoffice.gov.uk/climate-change/guide/impacts/high-end/map

Met Office (2011b) 2010 – a near-record year, 20 January 2011. Met Office. http://www.metoffice.gov.uk/news/releases/archive/2011/2010-global-temperature

NASA (2012) Visualizing the 2012 Sea Ice Minimum, NASA Earth Observatory, September 27, 2012. http://earthobservatory.nasa.gov/IOTD/view.php?id=79256

NRC (2010) *Advancing the Science of Climate Change.* National Research Council. ISBN-10: 0-309-14588-0.

NSIDC (2012) Media Advisory: Arctic sea ice breaks lowest extent on record. National Snow and Ice Data Centre, NSIDC 27 August. http://nsidc.org/news/press/20120827_2012extentbreaks2007record.html

Parmesan, C. and Yohe, G. (2002) 'A globally coherent fingerprint of climate change impacts across natural systems', *Nature* 421, pp. 37-42. DOI:10.1038/nature01286; 2 January. http://www.nature.com/nature/journal/v421/n6918/abs/nature01286.html

Peters, G. P. et al. (2013) 'The challenge to keep global warming below 2°C', *Nature Climate Change,* 3, January. http://www.nature.com/natureclimatechange

Richardson, Andrew D. et al. (2013) 'Climate change, phenology, and phenological control of vegetation feedbacks to the climate system', *Agricultural and Forest Meteorology,* 169, 156.

Rupp, D. E. et al. (2012) 'Did human influence on climate make the 2011 Texas drought more probable?' [in: Peterson, T. C., P. A. Stott, S. Herring, eds.], 'Explaining extreme events of 2011 from a climate perspective', *Bulletin of the American Meteorological Society* 93, pp. 1041-1067. DOI: 10.1175/BAMS-D-11-00021.1.

Schär, C. et al. (2004) 'The role of increasing temperature variability in European summer heatwaves', *Nature* 427(6972) pp. 332-336.

Schuur, Edward A. G. et al. (2008) Vulnerability of Permafrost Carbon to Climate Change: Implications for the Global Carbon Cycle, *BioScience,* 58(8):701-714., American Institute of Biological Sciences, DOI: http://dx.doi.org/10.1641/B580807 http://www.bioone.org/doi/full/10.1641/B580807

Shakun, Jeremy D. et al. (2012) 'Global warming preceded by increasing carbon dioxide concentrations during the last deglaciation', *Nature*, 484, pp. 49-54.

Shearer, C. and Rood, R. B. (2011) 'Changing the Media Discussion on Climate and Extreme Weather', *earthzine.org*, 17 April. http://www.earthzine.org/2011/04/17/changing-the-mediadiscussion-on-climate-and-extreme-weather

Tyndall, J. (1861) 'On the absorption and radiation of heat by gases and vapours, and on the physical connexion of radiation, absorption, and conduction', Bakerian lecture, *Philosophical Magazine Series 4*, 22(146) pp. 169-194. http://www.tandfonline.com/doi/abs/10.1080/14786446108643138

World Bank (2012) Turn Down the Heat: Why a 4°C Warmer World Must be Avoided, A Report for the World Bank by the Potsdam Institute for Climate Impact Research and Climate Analytics, November.

2.1.3 Planetary boundaries

Raworth, K. (2012) A Safe and Just Space for Humanity. Oxfam Discussion Paper, February.

Rockström, J. and M. Klum. (2012) The Human Quest: Prospering within Planetary Boundaries. [E-book] thehumnanquest.org

Rockström, J. (2010) Johan Rockstrom: Let the environment guide our development. TEDTalks, youtube.com, 31 August. http://www.youtube.com/watch?feature=player_embedded&v=RgqtrlixYR4

Rockström, J. et al. (2009) A safe operating space for humanity, *Nature* 461 (472-475), 24 September.

2.1.4 Future generations

Barnett, J. and W. Neil Adger (2007) Climate change, human security and violent conflict, *Political Geography, Climate Change and Conflict*, Volume 26, Issue 6, August 2007, pp. 639-655.

Beckerman, W. (1995) 'Small is Stupid: Blowing the whistle on the greens', *Population and Development Review*, 21 (3), September. London: Duckworth.

Brundtland, G.H. (1987) *Our Common Future*. Oxford University Press.

Nordhaus, W. D. (2007) 'A Review of the Stern Review on the Economics of Climate', *Journal of Economic Literature*, 45(3), pp. 686-702.

Sachs, Jeffrey D. (2007) Climate Change Refugees, Sustainable Developments, *Scientific American* 296, 43. doi:10.1038/scientificamerican0607-43

Stern, N. (2009) *A Blueprint for a Safer Planet: How to Manage Climate Change and Create a New Era of Progress and Prosperity*. Bodley Head.

Toman, M. (1998) 'Why not to calculate the value of the world's ecosystem services and natural capital', *Ecological Economics*, 25, pp. 57-60.

World Bank (2012) Turn Down the Heat: Why a 4°C Warmer World Must be Avoided, A Report for the World Bank by the Potsdam Institute for Climate Impact Research and Climate Analytics, November.

2.2 The situation in the long-industrialised West

2.2.1 Energy supplies

DECC (2009) Digest of United Kingdom Energy Statistics 2009. Department of Energy and Climate Change, London, The Stationery Office.

Johnson, V. et al. (2012) Economics of Oil Dependency: A glass ceiling to recovery. new economics foundation (nef). http://www.neweconomics.org/publications/the-economics-of-oil-dependence-a-glassceiling-to-recover.

Morgan, T. (2013) Perfect Storm – energy, finance and the end of growth. *Tullet Prebon, Strategy Insights: Issue Nine*, January. [PDF] http://www.tullettprebon.com/Documents/strategyinsights/TPSI_009_Perfect_Storm_009.pdf

Richards, P. (2012) Shale Gas and Fracking, UK Parliament Technical Briefing Papers No.SN06073.

2.2.2 The economic crisis

Elliot, Larry and Dan Atkinson (2012) *Going South*. Palgrave Macmillan. ISBN-10: 0230392547, ISBN-13: 978-0230392540.

2.2.3 Wellbeing

Abdallah, S. et al. (2012) The Happy Planet Index: 2012 Report. A global index of sustainable well-being. new economics foundation (nef), London.

Abdallah, S. et al. (2009) The Happy Planet Index 2.0: Why good lives don't have to cost the Earth. new economics foundation (nef) London.

Abdallah, S. et al. (2006) The Happy Planet Index: An index of human well-being & environmental impact, London: new economics foundation (nef).

Hills et al. (2010) An anatomy of economic inequality in the UK: Report of the National Equality Panel, Centre for Analysis of Social Exclusion (CASE) report 60. London: Government Equalities Office (GEO)/The London School of Economics and Political Science (LSE), London.

Lyubormirsky et al. (2005) 'Pursuing happiness: the architecture of sustainable change', *Review of General Psychology*, 9 (2), pp. 111–131.

Peston, R. (2009) 'Bosses' pay and WPP', *BBC: Peston's picks*, 2 June. http://www.bbc.co.uk/blogs/thereporters/robertpeston/2009/06/bosses_pay_and_wpp.html [Live at: June 2013].

Thompson, S. et al. (2007) The European (un)Happy Planet Index: An index of carbon efficiency and wellbeing in the EU. new economics foundation (nef), London.

Wilkinson, R. and K. Pickett (2009) *The Spirit Level: Why More Equal Societies Almost Always Do Better*. London: Allen Lane, London.

2.3 What does this mean for the UK?

Beales, C. (2013) UK emission targets. chrisbeales.net. http://www.chrisbeales.net/environment/emissions_UK%20targets.html [Accessed 25.5.13]

DECC (2011) The Carbon Plan: Delivering our low carbon future. Presented to Parliament pursuant to Sections 12 and 14 of the Climate Change Act 2008, Amended 2nd December 2011 from the version laid before Parliament on 1st Department of Energy and Climate Change, December.

HM Government (2008) Climate Change Act 2008, UK Policy Governance Association (UKPGA). HM Government, December. [PDF] http://www.legislation.gov.uk/ukpga/2008/27/pdfs/ukpga_20080027_en.pdf

Messner, D. et al. (2010) 'The Budget Approach: A Framework for a Global Transformation toward a Low-Carbon Economy', *Journal of Renewable and Sustainable Energy*, 2 (3).

UNFCCC (1998) The Kyoto Protocol to the United Nations Framework Convention on Climate Change. The United Nations Framework Convention on Climate Change (UNFCCC). http://unfccc.int/resource/docs/convkp/kpeng.pdf Accessed 20/05/13

2.3.1 Our carbon budget

DECC (2013a) Statistical Release: 2012 UK Greenhouse Gas Emission, Provisional Figures and 2011 UK Greenhouse Gas Emissions, Final Figures by Fuel Type and End-user, 28 March 2013. Department of Energy and Climate Change.

DECC (2013b) Statistical Release: 2011 UK Greenhouse Gas Emissions, Final Figures, 5 February 2013. Department of Energy and Climate Change.

FoE (2010) Reckless gamblers: How politicians' inaction is ramping up the risk of dangerous climate change – A report for policy makers by Friends of the Earth England, Wales & Northern Ireland, December.

Meinshausen, et al. (2009) 'Greenhouse-gas emission targets for limiting global warming to 2°C', *Nature* 458, pp. 1158-1162.

Messner, D. et al. (2010) 'The Budget Approach: A Framework for a Global Transformation toward a Low-Carbon Economy', *Journal of Renewable and Sustainable Energy*, 2 (3).

Schellnhuber, H.J. (2009) 'Four Degrees and Beyond', Presentation to International Climate Conference, 28-30 September 2009, Oxford.

Wei, T. et al. (2012) 'Developed and developing world responsibilities for historical climate change and CO_2 mitigation', *Proc. Natl. Acad. Sci.*, 109 (32) pp. 12911-12915.

Chapter 3. Our scenario: Rethinking the Future

3.1 About our scenario

3.1.2 Rules

Audsley, E. et al. (2009) How low can we go? An assessment of greenhouse gas emissions from the UK food system and the scope to reduce them by 2050. WWF-UK.

DECC (2013) Official statistics: Final UK greenhouse gas emissions, 5 February 2013. Department of Energy and Climate Change. https://www.gov.uk/government/publications/final-uk-emissions-estimates

DECC (2011) Management of the UK's plutonium stocks: A consultation response on the longterm management of UK-owned separated civil plutonium. Department of Energy and Climate Change, 1 December. [PDF] https://www.gov.uk/government/uploads/system/uploads/attachment_data/file/42773/3694-govt-resp-mgmt-of-uk-plutonium-stocks.pdf

IPCC (2005) IPCC Special Report on Carbon Dioxide Capture and Storage. Prepared by Working Group III of the Intergovernmental Panel on Climate Change [Metz, B., O. Davidson, H. C. de Coninck, M. Loos, and L. A. Meyer (eds.)]. Cambridge University Press, Cambridge, United Kingdom and NewYork, NY, USA, 442 pp.

Ipsis Mori (2010) Experiment Earth? Report on a Public Dialogue on Geoengineering, August.

Jenkins, G. J. et al. (2009). UK Climate Projections: Briefing report. Met Office Hadley Centre, Exeter, UK.

Lee, D.S. (2010) DfT Aviation Environment and Atmospheric Expert Technical Support. Department for Transport. [PDF] http://assets.dft.gov.uk/publications/pgr-aviation-research-researchreport-finalreport/DfT_FinalReport_250310.pdf

ONS (2012) UK Environmental Accounts, 2012, List of greenhouse gases with corresponding Global Warming Potential (GWP). Office of National Statistics (ONS). https://www.gov.uk/uk-greenhouse-gas-emissions#introduction

Royal Society (2009) Geoengineering the climate: Science, governance and uncertainty. The Royal Society, September.

UNEP (2012) The Emissions Gap Report 2012. United Nations Environment Programme (UNEP), Nairobi.

Vaughan, N. E. and T. M. Lenton (2011) A review of climate geoengineering proposals, Climatic Change, Volume 109, Issue 3-4, pp. 745-790, December.

3.1.3 Assumptions

Communities and Local Government (2010) Housing Statistical Release: Household Projections, 2008 to 2033, England, Communities and Local Government, 26 November 2010. [PDF] https://www.gov.uk/government/uploads/system/uploads/attachment_data/file/6395/1780763.pdf

Meadows, D. (1972) *The limits to growth*. Universe Books. ISBN: 0-87663-165-0.

ONS (2012) Labour Force Survey (LFS): Families and Households, 2001 to 2011, Demographic Analysis Unit, Office for National Statistics, 19 January.

ONS (2011) National Population Projections – ONS 2010-based, 26 October.

Utley, J. and Shorrock, L. (2008) Domestic Energy Fact File 2008, BRE Housing and the Department of Energy and Climate Change. [PDF] http://www.bre.co.uk/filelibrary/pdf/rpts/Fact_File_2008.pdf

3.2 Measuring up today

Bain, C.G. et al. (2011) IUCN UK Commission of Inquiry on Peatlands. IUCN UK Peatland Programme, Edinburgh. Table 1 Summary of organic-rich soils extent and bogs and fen UK BAP type extent; adapted with kind permission from JNCC (2011).

DECC (2013) Official statistics: Final UK greenhouse gas emissions, 5 February 2013. Department of Energy and Climate Change. https://www.gov.uk/government/publications/final-uk-emissions-estimates

DECC (2012a) Digest of United Kingdom Energy Statistics 2012, Department of Energy and Climate Change, 26 July 2012. https://www.gov.uk/government/organisations/department-of-energy-climate-change/series/digest-of-uk-energy-statistics-dukes

DECC (2012b) Energy consumption in the UK, Department of Energy and Climate Change, 26 July 201. www.gov.uk/government/publications/energy-consumption-in-the-uk

DEFRA (2012) June Surveys Census of Agriculture SAF land data Scotland, 07 June 2012; Table 3.1 Agricultural land use (a). http://www.defra.gov.uk/statistics/foodfarm/cross-cutting/auk/

Forestry Commission (2007) Forestry Facts & Figures, 2007, A summary of statistics about woodland and forestry. Forestry Commission, 2007. [PDF] http://www.forestry.gov.uk/pdf/fcfs207.pdf/$FILE/fcfs207.pdf

Morton, D. et al. (2008) CS Technical Report No 11/07: Final Report for LCM2007 - the new UK Land Cover Map. Centre for Ecology & Hydrology, (Natural Environment Research Council), July 2. [PDF] http://www.ceh.ac.uk/documents/LCM2007FinalReport.pdf

NERC (2008) Countryside Survey: UK Results from 2007 Chapter 2 – The National Picture. Natural Environment Research Council, November 2008; Table 2.3: Estimated area ('000s ha) of selected Priority Habitats in Great Britain in 1998 and 2007. Estimates for 1998 could not be calculated for all Priority Habitats.

Read, D.J. et al. (2009) Combating climate change – a role for UK forests. An assessment of the potential of the UK's trees and woodlands to mitigate and adapt to climate change. The Stationery Office, Edinburgh.

3.3 Power Down

DECC (2013) Official statistics: Final UK greenhouse gas emissions, 5 February 2013. Department of Energy and Climate Change. https://www.gov.uk/government/publications/final-uk-emissions-estimates

DECC (2012) Energy consumption in the UK, Department of Energy and Climate Change, 26 July 2012. www.gov.uk/government/publications/energy-consumption-in-the-uk

3.3.1 Buildings and industry

DECC (2013) Official statistics: Final UK greenhouse gas emissions, 5 February 2013. Department of Energy and Climate Change. https://www.gov.uk/government/publications/final-uk-emissions-estimates

DECC (2012) Energy consumption in the UK. Department of Energy and Climate Change, 26 July 2012. https://www.gov.uk/government/publications/energy-consumption-in-the-uk

DECC (2010) 2050 Pathways Analysis, July 2010. Department of Energy and Climate Change.

DEFRA (2010) UK's Carbon Footprint 1993-2010, 13 December 2012. Department for Environment, Food and Rural Affairs. [PDF] https://www.gov.uk/government/uploads/system/uploads/attachment_data/file/85869/release-carbon-footprint-dec2012.pdf

NERA (2010) Decarbonising Heat: Low-Carbon Heat Scenarios for the 2020s, June 2010. NERA Economic Consulting and AEA.

3.3.2 Transport

CCC (2012) Meeting Carbon Budgets – 2012 Progress Report to Parliament, 1 June 2012. Committee on Climate Change. http://www.theccc.org.uk/publications

DECC (2013) Official statistics: Final UK greenhouse gas emissions, 5 February 2013. Department of Energy and Climate Change. https://www.gov.uk/government/publications/final-uk-emissions-estimates

DECC (2012) Energy consumption in the UK. Department of Energy and Climate Change, 26 July 2012. https://www.gov.uk/government/publications/energy-consumption-in-the-uk

DECC (2010) 2050 Pathways Analysis, July 2010. Department of Energy and Climate Change.

DfT (2012) Transport Statistics Great Britain 2012, 13 December 2012. Department for Transport. https://www.gov.uk/government/publications/transport-statistics-great-britain-2012

DfT (2009a) Transport Statistics Bulletin – National Travel Survey: 2008. Department for Transport.

DfT (2009b) UK Aviation: Carbon Reduction Futures, July 2009. Department for Transport. [PDF] http://assets.dft.gov.uk/publications/pgr-aviation-environmentalissues-carbonreductionfutures/finalreport.pdf

Lee, D.S. (2010) DfT Aviation Environment and Atmospheric Expert Technical Support. Department for Transport. [PDF] http://assets.dft.gov.uk/publications/pgr-aviation-research-researchreport-finalreport/DfT_FinalReport_250310.pdf

3.4 Power Up

DECC (2012) Digest of United Kingdom Energy Statistics 2012, August 2012. Department of Energy and Climate Change. https://www.gov.uk/government/organisations/department-of-energy-climate-change/series/digest-of-uk-energy-statistics-dukes

3.4.1 Renewable energy supply

Arup (2011) Review of the generation costs and deployment potential of renewable electricity technologies in the UK – A report for the Department of Energy and Climate Change, October 2011. Ove Arup & Partners Ltd. [PDF] http://www.decc.gov.uk/assets/decc/11/consultation/ro-banding/3237-cons-ro-banding-arup-report.pdf

DECC (2013) Official statistics: Final UK greenhouse gas emissions, 5 February 2013. Department of Energy and Climate Change. https://www.gov.uk/government/publications/final-uk-emissions-estimates

DECC (2012) Digest of United Kingdom Energy Statistics 2012, August 2012. Department of Energy and Climate Change. https://www.gov.uk/government/organisations/department-of-energy-climate-change/series/digest-of-uk-energy-statistics-dukes

DECC (2010) 2050 Pathways Analysis, July 2010. Department of Energy and Climate Change. https://www.gov.uk/2050-pathways-analysis

Offshore Valuation Group (2010) A valuation of the UK's offshore renewable energy resource, May 2010. Public Interest Research Centre (PIRC). http://www.offshorevaluation.org

Troen, I. and E.L. Petersen (1989). European Wind Atlas. ISBN 87-550-1482-8. Risø National Laboratory, Roskilde. 656 pp.

Pöyry (2011) Analysing Technical Constraints on Renewable Generation to 2050 – A report to the Committee on Climate Change, March 2011. Pöyry Management Consulting. [PDF] http://hmccc.s3.amazonaws.com/Renewables%20Review/232_Report_Analysing%20the%20technical%20constraints%20on%20renewable%20generation_v8_0.pdf

3.4.2 Balancing supply and demand

GridGas (2012) Power To Gas. GridGas consortium, ITM Power Plc, 2012. http://www.gridgas.co.uk/power-to-gas.html

3.4.3 Transport and industrial fuels

Hughes, A. D. et al. (2012) 'Biogas from Macroalgae: is it time to revisit the idea?', *Biotechnology for Biofuels 2012*, 5:86.

Kumar, A. at el. (2011) 'Enhanced CO_2 fixation and biofuel production via microalgae: recent developments and future directions', *Trends in Biotechnology* 28(7).

Roberts, T., and Paul U. (2012) 'Prospects for the use of macro-algae for fuel in Ireland and the UK: An overview of marine management issues', *Marine Policy* 36 (2012), pp. 1047-1053.

3.5 Non-energy emissions

3.5.1 Industry, businesses and households

AEA (2010) Analysing the Opportunities for Abatement in Major Emitting Industrial Sectors, 8th December 2010. AEA.

DECC (2013) Official statistics: Final UK greenhouse gas emissions, 5 February 2013. Department of Energy and Climate Change. https://www.gov.uk/government/publications/final-uk-emissions-estimates

DEFRA (2013) 'National Atmospheric Emissions Inventory: UK emissions data selector'. Department for Environment, Farming and Rural Affairs. http://naei.defra.gov.uk/data/data-selector

Lucas, P. et al. (2007) 'Long-term reduction potential of non-CO_2 greenhouse gases'. *Environmental Science & Policy*, 10(2), pp. 85-103.

ONS (2012) UK Environmental Accounts, 2012, List of greenhouse gases with corresponding Global Warming Potential (GWP), Office of National Statistics (ONS), https://www.gov.uk/uk-greenhouse-gas-emissions#introduction

ULCOS (Ultra-Low Carbon Dioxide Steelmaking) (2010a) 'Sustainable Biomass'. ULCOS. http://www.ulcos.org/en/research/substainable_biomass.php

ULCOS (Ultra-Low Carbon Dioxide Steelmaking) (2010b) 'Alkaline Electrolysis'. ULCOS. http://www.ulcos.org/en/research/electrolysis.php

3.5.2 Waste

AEA Technology Environment (1998a) Options to reduce methane emissions (Final Report), A report produced for DGXI, November. AEA Technology Environment. [PDF] http://ec.europa.eu/environment/enveco/climate_change/pdf/methane_emissions.pdf

AEA Technology Environment (1998b) Options to Reduce Nitrous Oxide Emissions (Final Report), A report produced for DGXI, November. AEA Technology Environment. [PDF] http://ec.europa.eu/environment/enveco/climate_change/pdf/nitrous_oxide_emissions.pdf

DECC (2013) Official statistics: Final UK greenhouse gas emissions, 5 February 2013. Department of Energy and Climate Change. https://www.gov.uk/government/publications/final-uk-emissions-estimates

DEFRA (2011a) ENV23 – UK waste data and management, Waste and recycling statistics, 4 August 2011. Department for Environment Food & Rural Affairs. https://www.gov.uk/government/organisations/department-for-environment-food-rural-affairs/series/waste-and-recycling-statistics

DEFRA (2011b) Government Review of Waste Policy in England. Department for Environment Food & Rural Affairs.

Environment Agency (undated) Limiting Climate Change. The Environment Agency.

FAO (2011) Global Food Losses and Food Waste: Extent, Causes and Prevention. Food and Agriculture Organization of the United Nations, Rome, Italy.

Fawcett, et al. (2002) Carbon UK: ECI RESEARCH REPORT 25, March. Industrial Sustainable Development Group Environmental Change Institute, University of Oxford.

Hogg, D. et al. (2011) Inventory Improvement Project – UK Landfill Methane Emissions Model, Final Report to Defra, January. Eunomia Research & Consulting Ltd.

Michaud, J. C. et al. (2010) Environmental benefits of recycling – 2010 update, March 2010. Copenhagen Resource Institute.

Symonds, in association with ARGUS, COWI and PRC Bouwcentrum (1999) Construction and Demolition Waste Management Practices and their Economic Impacts (Final Report), A report produced for DGXI, February. European Commission. [PDF] http://ec.europa.eu/environment/waste/studies/cdw/cdw_chapter1-6.pdf

UNEP (2010) Waste and Climate Change: Global trends and strategy framework. United Nations Environmental Programme Division of Technology, Industry and Economics International Environmental Technology Centre Osaka/Shiga, United Nations Environment Programme.

WRAP (2013) Facts and figures, Waste & Resources Action Programme. http://www.wrap.org.uk/content/facts-and-figures [Live at: March 2013]

3.6 Land use

3.6.1 Agriculture, food and diets

Audsley, E. et al. (2009) How low can we go? An assessment of greenhouse gas emissions from the UK foodsystem and the scope to reduce them by 2050. WWF-UK.

Akiyama, H. et al. (2010) Evaluation of effectiveness of enhanced-efficiency fertilizers as mitigation options for N_2O and NO emissions from agricultural soils: meta-analysis, *Global Change Biology*, 16, pp. 1837-1846.

Bates, B. et al. (2011) National Diet and Nutrition Survey: Headline results from Years 1, 2 and 3 (combined) of the Rolling Programme *(2008/2009 – 2010/11)*. Department of Health and the Food Standards Agency. London, UK.

Bhat, Z. F. and Bhat, H. (2011) 'Tissue engineered meat – future meat', *Journal of Stored Products and Postharvest Research*, 2(1), pp. 1-10. http://www.academicjournals.org/JSPPR

Blengini, G. A. and Busto, M. (2009) 'The life cycle of rice: LCA of alternative agri-food chain management systems in Vercelli (Italy)', *Journal of Environmental Management*, 90, pp. 1512-1522.

Buddle, B. M. et al. (2011) 'Strategies to reduce methane emissions from farmed ruminants grazing on pasture', *The Veterinary Journal*, 188 (1), pp. 11-17. http://dx.doi.org/10.1016/j.tvjl.2010.02.019

COMA (1991) Dietary Reference Values (DRVs) for Food Energy and Nutrients for the UK. Committee on Medical Aspects of Food Policy (COMA). The Stationary Office, London, UK.

Datar, I. and Betti, M. (2009) 'Possibilities for an in vitro meat production system', *Innovative Food Science and Emerging Technologies*, 11, pp.13-22.

DECC (2012) Alternative approaches to reporting UK greenhouse gas emissions. Department of Energy & Climate Change. London, UK

DEFRA (2012) June Surveys Census of Agriculture SAF land data Scotland, 07 June 2012; Table 3.1 Agricultural land use (a). http://www.defra.gov.uk/statistics/foodfarm/cross-cutting/auk/

DEFRA (2011) Agriculture in the United Kingdom 2011. Department for Environment, Food and Rural Affairs, York, UK.

Devos, Y. et al. (2009) 'Coexistence of genetically modified (GM) and non-GM crops in the European Union. A review', *Agronomy for Sustainable Development*, 29 (1), pp. 11-30.

Di, H.J. and Cameron, K, C. (2012) 'How does the application of different nitrification inhibitors affect nitrous oxide emissions and nitrate leaching from cow urine in grazed pastures?', *Soil Use and Management*, 28 (March), pp.54-61.

Eckard, R. J. et al. (2010) 'Options for the abatement of methane and nitrous oxide from ruminant production: A review', *Livestock Science*, 130 (1-3), pp.47-56. http://dx.doi.org/10.1016/j.livsci.2010.02.010

FAO (2011) Global Food Losses and Food Waste: Extent, Causes and Prevention. Food and Agriculture Organization of the United Nations, Rome, Italy.

Friel, S. et al. (2009) 'Public health benefits of strategies to reduce greenhouse-gas emissions: food and agriculture', *The Lancet* 374, pp. 2016-2025.

FSA (2007) FSA nutrient and food based guidelines for UK institutions. The Food Standards Agency and the Department of Health, London, UK.

Garnett, T. (2007) Meat and Dairy Production and Consumption. Food Climate Research Network, University of Surrey, UK

Godfray, H.C. J. et al. (2010) 'Food security: the challenge of feeding 9 billion people', *Science* (New York, N.Y.), 327 (5967), pp. 812-8.

González, A. D. et al. (2011) 'Protein efficiency per unit energy and per unit greenhouse gas emissions: Potential contribution of diet choices to climate change mitigation', *Food Policy*, 36 (5) 562.

Grainger, C. and Beauchemin, K. A. (2011) 'Can enteric methane emissions from ruminants be lowered without lowering their production?', *Animal Feed Science and Technology*, 166-167, pp. 308-320. http://dx.doi.org/10.1016/j.anifeedsci.2011.04.021

Holding, J. et al. (2011) Food Statistics Pocketbook 2011. Food Statistics Branch: Department for Environment, Food and Rural Affairs. York, UK.

Hongmin, D. et al. (2011) 'Reducing methane production from livestock: can more efficient livestock systems help?', Development of sustainable livestock systems on grasslands in north-western China. Austrailian Centre for International Agricultural Research, Australia.

Lang, T. and Rayner, G. (2007) 'Overcoming policy cacophony on obesity: an ecological public health framework for policymakers', *Obesity reviews: an official journal of the International Association for the Study of Obesity*, 8 (Suppl 1), pp. 165-81.

Liu, C. et al. (2013) Effects of nitrification inhibitors (DCD and DMPP) on nitrous oxide emission, crop yield and nitrogen uptake in a wheat-maize cropping system', *Biogeosciences Discuss*, 10, pp.711-737.

McMichael, A.J. et al. (2007) 'Food, livestock production, energy, climate change, and health', *Lancet*, 370 (9594), pp. 1253-63.

Millstone, E. and Lang, T. (2008) *The Atlas of Food*, Second Edition. Earthscan, Brighton.

Miraglia, M. et al. (2009) 'Climate change and food safety: An emerging issue with special focus on Europe', *Food and Chemical Toxicology* 47, pp. 1009-1021.

Mironov, V. et al. (2009) 'Biofabrication: a 21st century manufacturing paradigm', *Biofabrication*, 1, pp. 1-16.

Monsivais, P. and Drewnowski, A. (2007) 'The Rising Cost of Low-Energy-Density Foods', *Journal of American Dietetics Association*, 107, pp. 2071-2076.

Pan, A. et al. (2012) 'Red Meat Consumption and Mortality: Results from 2 Prospective Cohort Studies', *Archives of International Medicine*, 7, pp. 1-9.

Poskitt, E.M.E. (2009) 'Countries in transition: underweight to obesity non-stop?' *Annals of tropical paediatrics*, 29 (1), pp. 1-11.

Post, M. J. (2012) Cultured meat from stem cells: Challenges and prospects. *Meat Science* 92 pp. 297–301.

Sinclair Knight Merz (2012) Geothermal Energy Potential: Great Britain and Northern Ireland. Sinclair Knight Merz, London, UK.

Stockley, L. et al. (2007) Nutrient profiles for use in relation to food promotion and children's diet: Update of 2004 literature review. University of Oxford, Oxford, UK.

Tuomisto, H. L. and de Mattos, M. J. T. (2010) Life cycle assessment of cultured meat production, 7th International Conference on Life Cycle Assessment in the Agri-Food Sector. Bari, Italy.

War on Want (2012) Food Sovereignty: Reclaiming the global food system. War on Want, London, UK.

WHO (2013) Non-Communicable Diseases: Country Profiles. World Health Organization, Geneva, Switzerland.

WHO (2003) Fruit and Vegetable Promotion Initiative. World Health Organisation, Geneva, Switzerland.

3.6.2 Growing energy and fuel

Biomass Energy Centre (2011) Energy crops, Biomass Energy Centre http://www.biomassenergycentre.org.uk/portal/page?_pageid=75,17301&_dad=portal&_schema=PORTAL [Live at: 31 May 2013]

DEFRA (2012) June Surveys Census of Agriculture SAF land data Scotland, 07 June 2012; Table 3.1 Agricultural land use (a). http://www.defra.gov.uk/statistics/foodfarm/cross-cutting/auk/

DEFRA (2009) Science and Research Projects: The Genetic Improvement of Miscanthus as a Sustainable Feedstock for Bioenergy in the UK - NF0446. Department for Environment, Food and Rural Affairs.

Haughton, A.J. et al. (2009) 'A novel, integrated approach to assessing social, economic and environmental implications of changing rural land use: a case study for perennial biomass crops', *Journal of Applied Ecology*, 46(2), pp. 315-322.

Sims, R.E.H. et al. (2006) 'Energy crops: current status and future prospects', *Global Change Biology* 5, 12 pp. 2054-76.

3.6.3 Capturing carbon

Andres, R. J. et al. (1999) Carbon dioxide emissions from fossil-fuel use, pp. 1751-1950, *Tellus* (1999), 51B, 28 December 1998, pp. 759-765. ISSN 0280–6495.

Atkinson, S. and M. Townsend (2011) The State of the UK's Forests, Woods and Trees: Perspectives from the sector, A report to mark the International Year of Forests 2011. Woodland Trust.

Augenstein, D. (2001) Landfill Operation for Carbon Sequestration and Maximum Methane Emission Control, Controlled Landfilling Demonstration Cell Performance for Carbon Sequestration, Greenhouse Gas Emission Abatement and Landfill Methane Energy. Final Report, Reporting Period: July 1, 1999 to February 27, 2001.

Bain, C. G. et al. (2011) IUCN UK Commission of Inquiry on Peatlands. IUCN UK Peatland Programme, Edinburgh.

Bell, M. J. et al. (2011) 'UK land-use change and its impact on SOC: 1925–2007', *Global Biogeochem*, Cycles 25, GB4015.

Bellarby, J. et al. (2013) 'Livestock greenhouse gas emissions and mitigation potential in Europe', *Global Change Biology* 19, pp. 3-18.

Broadmeadow, M. and R. Matthews (2003) INFORMATION NOTE: Forests, Carbon and Climate Change: the UK Contribution, Forest Research, June. Forestry Commission.

Dawson, J.J.C. and Smith, P. (2007) 'Carbon losses from soil and its consequences for land-use management', *Science of the Total Environment* 382, pp. 165-190.

DECC (2013) Official statistics: Final UK greenhouse gas emissions, 5 February 2013. Department of Energy and Climate Change. https://www.gov.uk/government/publications/final-uk-emissions-estimates

Dewar, R. C. and Cannell, M. G. R. (1991) Carbon sequestration in the trees, products and soils of forest plantations: an analysis using UK examples, December 12, 1991, *Tree Physiology* 11, 49-71.

Fawcett, T. et al. (2002) Carbon UK: ECI RESEARCH REPORT 25, March. Industrial Sustainable Development Group Environmental Change Institute, University of Oxford.

Forestry Commission (2010) Wood fibre availability and demand in Britain 2007 to 2025, Prepared for Confederation of Forest Industries (Confor) UK Forest Products Association (UKFPA) Wood Panel Industries Federation (WPIF), by John Clegg Consulting Ltd, March 2010. The Forestry Commission.

Groenigen, K. J. van et al. (2011) 'Soil C storage as affected by tillage and straw management: An assessment using field measurements and model predictions', *Agriculture, Ecosystems & Environment* Volume 140, Issues 1–2, 30 January 2011, pp. 218-225. http://www.sciencedirect.com/science/article/pii/S0167880910003233

Hammond, J. et al. (2011) Prospective life cycle carbon abatement for pyrolysis biochar systems in the UK, *Energy Policy* 39 (2011), pp. 2646-2655.

IPCC (2007) Summary for policy makers. Contribution of Working Group III to the Fourth Assessment Report of the Intergovernmental Panel on Climate Change. In: Metz, O.R. Davidson, P.R. Bosch, R. Dave and L.A. Meyer (eds) Climate Change 2007: Mitigation, Cambridge University Press, Cambridge, UK and New York, NY, USA.

IPCC (2000) IPCC Good Practice Guidance and Uncertainty Management in National Greenhouse Gas Inventories 5.1, Chapter 5 Waste. Intergovernmental Panel on Climate Change, accepted at the IPCC Plenary at its 16th session held in Montreal, 1-8 May, 2000. http://www.ipcc-nggip.iges.or.jp/public/gp/english

Littlewood, N. et al. (2010) Peatland biodiversity. Report to IUCN UK Peatland Programme, Edinburgh. www.iucn-uk-peatlandprogramme.org/scientificreviews

McCarl, B. A. and R. D. Sands (2007) 'Competitiveness of terrestrial greenhouse gas offsets: are they a bridge to the future?', *Climate Change*, Volume 80, Numbers 1-2 (2007), pp. 109-126. DOI: 10.1007/s10584-006-9168-5. http://www.springerlink.com/content/c6734vxht0m41331

Moran, D. et al. (2011) 'Developing carbon budgets for UK agriculture, land-use, land-use change and forestry out to 2022', *Climatic Change*, Volume 105, Numbers 3-4 (2011), pp. 529-553. DOI: 10.1007/s10584-010-9898-2. http://www.springerlink.com/content/046060375771462g

Morison, J. et al. (2012) Understanding the carbon and greenhouse gas balance of forests in Britain, Forestry Commission Research Report. Forestry Commission, Edinburgh. i–vi + pp. 1-149

Ostle, N. J. et al. (2009) UK land use and soil carbon sequestration, *Land Use Policy* Volume 26, Supplement 1, December, pp. S274-S283. http://www.sciencedirect.com/science/article/pii/S0264837709000945

Parliamentary Office of Science and Technology (2010) HoP POSTNOTE Number 358 July 2010: Biochar. The Parliamentary Office of Science and Technology. [PDF] http://www.parliament.uk/documents/post/postpn358-biochar.pdf

Read, D. J. et al. (eds). (2009) Combating climate change – a role for UK forests. An assessment of the potential of the UK's trees and woodlands to mitigate and adapt to climate change. The Stationery Office, Edinburgh.

Sadler, P. and D. Robson (undated) Carbon Sequestration by Buildings. The Alliance for Sustainable Building Products.

Shackley, S. and S. Sohi (eds) (undated) An Assessment of the Benefits and Issues Associated with the Application of Biochar to Soil, A report commissioned by the United Kingdom Department for Environment, Food and Rural Affairs, and Department of Energy and Climate Change. UK Biochar Research Centre. [PDF] http://www.geos.ed.ac.uk/homes/sshackle/SP0576_final_report.pdf

Shackley, S. et al. (2011) 'The feasibility and costs of biochar deployment in the UK', *Carbon Management* (2011) 2(3), pp. 335-356.

Smith, P. et al. (2010) 'Opportunities and risks of land management for soil carbon sequestration', Presentation at Food Climate Research Network, Soil carbon sequestration workshop, 21 January 2010. Nobel House.

Smith, P. (2008) 'Commentary: Do agricultural and forestry carbon offset schemes encourage sustainable climate solutions?', *International Journal of Agricultural Sustainability* 6 (3) 2008, pp. 169-170. Earthscan. DOI:10.3763/ijas. 2008. c5002, ISSN: 1473-5903 (print), 1747-762X www.earthscanjournals.com

Smith, P. (2008) 'Greenhouse gas mitigation in agriculture'. *Philosophical Transactions of the Royal Society B: Biological Sciences* 363, pp. 789-813.

Smith, P. et al. (2000) 'Meeting the UK's climate change commitments: options for carbon mitigation on agricultural land', *Soil Use and Management* 16, pp. 1-11.

Sohi, S.P. (2012) Carbon storage with benefits, *Science*, 338, pp. 1034-1035.

Sohi, S. P. et al. (2010) A Review of Biochar and its Use and Function in Soil. In Sparks, Donald L. (ed.), *Advances in Agronomy, Vol. 105*, pp. 47-82. Academic Press, Burlington. ISBN: 978-0-12-381023-6.

Verheijen, F. G. A. et al. (2009) Biochar Application to Soils – A Critical Scientific Review of Effects on Soil Properties, Processes and Functions. EUR 24099 EN, Office for the Official Publications of the European, Communities, Luxembourg. [PDF] http://eusoils.jrc.ec.europa.eu/esdb_archive/eusoils_docs/other/eur24099.pdf

West, T. O. and J. Six (2007) 'Considering the influence of sequestration duration and carbon saturation on estimates of soil carbon capacity', *Climatic Change*, Volume 80, Numbers 1-2 (2007), pp. 25-41. DOI: 10.1007/s10584-006-9173-8. http://www.springerlink.com/content/v637652820450657

Worrall, F. et al. (2011) A review of current evidence on carbon fluxes and greenhouse gas emissions from UK peatlands, JNCC research report 442. Peterborough. [PDF] http://jncc.defra.gov.uk/pdf/jncc442_webFinal.pdf

Zeng, N. (2008) 'Carbon sequestration via wood burial', *Carbon Balance and Management 2008,* 3:1, 3 January 2008. DOI:10.1186/1750-0680-3-1. http://www.cbmjournal.com/content/3/1/1

3.8 How we get there

3.8.1 ZCB and the UK's carbon budget

Anderson, K. and Bows, A. (2010) 'Beyond 'dangerous' climate change: emission scenarios for a new world', 29 November 2010, *Philosophical Transactions of the Royal Academy of Science* 369(1934), pp. 20-44, 13 January. DOI:10.1098/rsta.2010.0290.http://rsta.royalsocietypublishing.org/content/369/1934/20

Beales, C. (2013) UK emission targets, chrisbeales.net http://www.chrisbeales.net/environment/emissions_UK%20targets.html [Live at: 25.5.13]

Chichilnisky, G. (1994) North-South Trade and the Global Environment. *American Economic Review*, 84, 851.

DECC (2013a) Statistical Release: 2012 UK Greenhouse Gas Emission, Provisional Figures and 2011 UK Greenhouse Gas Emissions, Final Figures by Fuel Type and End-user, 28 March 2013. Department of Energy and Climate Change.

DECC (2013b) Statistical Release 2011 UK Greenhouse Gas Emissions, Final Figures, 5 February 2013. Department of Energy and Climate Change.

FoE (2010) Reckless gamblers: How politicians' inaction is ramping up the risk of dangerous climate change – A report for policy makers by Friends of the Earth England, Wales & Northern Ireland, December 2010.

UNEP (2012) The Emissions Gap Report 2012, United Nations Environment Programme (UNEP), Nairobi

3.8.2 Zero carbon policy

Carbon Trust (2010) Tackling carbon leakage: sector-specific solutions for a world of unequal carbon prices, March 2010 http://www.carbontrust.com/media/84908/ctc767-tackling-carbon-leakage.pdf

EREC (2010) Re-thinking 2050 – a 100% renewable energy vision for the European Union, April 2010. European Renewable Energy Council (EREC) [PDF] http://www.erec.org/fileadmin/erec_docs/Documents/Publications/ReThinking2050_full%20version_final.pdf

Fleming, D. Chamberlin, S. (2011) Tradable Energy Quotas: A Policy Framework for Peak Oil and Climate Change, January 2011. http://www.teqs.net/download

Roberts, S. and Thumin, J. (2006) A rough guide to individual carbon trading, the ideas, the issues and the next steps, report to DEFRA (Department for Food, Environment and Rural Affairs). [PDF] http://cse.375digital.co.uk/downloads/file/publ067.pdf

UNEP (2012) The Emissions Gap Report 2012, United Nations Environment Programme (UNEP), Nairobi.

White, V. et al. (2013) Personal Carbon Allowances: distributional implications associated with personal travel and opportunities to reduce household emissions, Project Paper No.3, March 2013. http://www.cse.org.uk/downloads/file/project_paper_3_personal_carbon_allowances.pdf

3.8.3 Economic transition

Green New Deal Group (2008) Green New Deal. new economics foundation (nef), London, July.

3.9 Benefits beyond carbon

3.9.1 Adaptation

Anderson, K. and Bows, A. (2010) Beyond 'dangerous' climate change: emission scenarios for a new world, 29 November 2010, *Philosophical Transactions of the Royal Academy of Science* 369(1934) pp. 20-44, 13 January 2011. doi:10.1098/rsta.2010.0290 http://rsta.royalsocietypublishing.org/content/369/1934/20

Atkinson, S. and M. Townsend (2011) The State of the UK's Forests, Woods and Trees: Perspectives from the sector, A report to mark the International Year of Forests 2011. Woodland Trust.

Bain, C. G. et al. (2011) IUCN UK Commission of Inquiry on Peatlands. IUCN UK Peatland Programme, Edinburgh.

DEFRA (2010) Adapting to climate change: A guide for local councils. Department for Environment, Food and Rural Affairs, January. https://www.gov.uk/government/uploads/system/uploads/attachment_data/file/166272/adapt-localcouncilguide.pdf.pdf [Accessed: 26.5.13]

Deser, C. et al. (2012) Communication of the role of natural variability in future North American climate, *Nature Climate Change*, vol. 2, p. 775-779. http://dx.doi.org/10.1038/nclimate1562.
Greenpeace (2007) The impacts of climate change on nuclear power station sites: a review of four proposed new-build sites on the UK coastline. Greenpeace.

HM Government (2012) UK Climate Change Risk Assessment: Government Report. In addition to this Government Report, the UK Climate Change Risk Assessment 2012 Evidence Report, which sets out the evidence base for the risk assessment, was laid before Parliament on 25 January 2012. Presented to Parliament pursuant to Section 56 of the Climate Change Act 2008, January 2012. The Stationery Office, London.

Jenkins, G. J. et al. (2009) UK Climate Projections: Briefing report. Met Office Hadley Centre, Exeter, UK.

NCCARF (2010) Climate Change Adaptation, National Climate Change Adaptation Research Facility, http://www.nccarf.edu.au/content/climate-change-adaptation [Accessed: 26 June 2012]

Read, D.J. et al. (2009) Combating climate change – a role for UK forests. An assessment of the potential of the UK's trees and woodlands to mitigate and adapt to climate change. The Stationery Office, Edinburgh.

Sohi, S. P. et al. (2010) A Review of Biochar and its Use and Function in Soil. In Sparks, Donald L. (ed.), *Advances in Agronomy, Vol. 105*, pp. 47-82. Academic Press, Burlington. ISBN: 978-0-12-381023-6.

3.9.2 Planetary boundaries

Elferink, E.V. et al. (2008) Feeding livestock food residue and the consequences for the environmental impact of meat, *J. Cleaner Production* 16 (12), pp. 1227-1233.

Foley, J.A., et al. (2011) 'Solutions for a cultivated planet', *Nature* 478, pp. 337-342.

Greenpeace International (2012) Ecological Livestock: Options for reducing livestock production and consumption to fit within ecological limits. Greenpeace Research Laboratories Technical Report 03/2012.

Haughton A. J. et al. (2009) 'A novel, integrated approach to assessing social, economic and environmental implications of changing rural land use', *J. Applied Ecology*, 46 (2) p. 315.

Hoekstra, A.Y. (2013) *The Water Footprint of Modern Consumer Society*. Routledge.

Lilywhite, R. and C.R. Rahn (2005) *Nitrogen UK*. Mass Balance Series, Biffa Ltd.

Lutz, W. et al. (eds) (2004) *The End of Population Growth in the 21st Century*. Earthscan.

Mekonnen, M.M. and A.Y. Hoekstra (2012) 'A Global Assessment of the Water Footprint of Farm Animal Products', *Ecosystems* **15**:401-415. DOI: 10.1007/s10021-011-9517-8

Nelleman, C. et al. (eds) (2009) The Environmental Food Crisis. UNEP.

Pelletier, N. and P. Tyedmers (2010) Forecasting potential global environmental costs of livestock production 2000–2050. http://www.pnas.org/cgi/doi/10.1073/pnas.1004659107

Rockström, J. and M. Klum. (2012). The Human Quest: Prospering within Planetary Boundaries. E-book: thehumanquest.org.

Rockström, J. et al. (2009) 'A safe operating space for humanity', *Nature* 461, 24 September 2009, pp. 472-475.

Sutton, M.A. et al. (2013) Our Nutrient World: The challenge to produce more food and energy with less pollution. Global Overview of Nutrient Management. Centre for Ecology and Hydrology, Edinburgh.

UNEP (2012) Handbook for the Montreal Protocol on Substances that Deplete the Ozone Layer. 9th Edition. United Nations Environment Programme.

Victor, P. (2008) *Managing without Growth*. Edward Elgar.

Williamson, P. and C. Turley (2012) 'Ocean acidification in a geoengineering context', *Phil. Trans. R. Soc. A*, 370 (1974), 13 September 2012, pp. 4317-4342.

3.9.3 Employment

DECC (2012) Renewable Energy Roadmap 2012 Update. Department of Energy and Climate Change (DECC).

Forestry Commission (2012) Forestry Statistics 2012, Economics & Statistics, Forestry Commission, National Statistics. www.forestry.gov.uk/statistics

Kemp, M. and Wexler, J. (2010) *ZeroCarbonBritain 2030: A new energy strategy*, Centre for Alternative Technology Publications.

Morris, C. and Pehnt, M (2012) German Energy Transition – Arguments for a Renewable Energy Future, November. Heinrich Böll Foundation. http://www.energytransition.de

National Careers Service (2012) Job market information: Finding out about agricultural livestock. National Careers Service, produced Sept 2010 using Lantra AACS LMI report, June 2010. https://nationalcareersservice.direct.gov.uk/advice/planning/LMI/Pages/agriculturallivestock.aspx

REA (2012) Renewable Energy: Made in Britain. Renewable Energy Association (REA).

RUK (2011) Working for a Green Britain. Renewable UK (RUK).

3.9.4 Wellbeing – measuring what matters

Abdallah, S. et al. (2012) The Happy Planet Index: 2012 Report. A global index of sustainable well-being. new economics foundation (nef), London.

Aked, J. and Thompson, S. (2011) *Five Ways to Wellbeing.* new economics foundation (nef).

3.10 Other scenarios

3.10.1 Scenario variations using ZCB rules

DECC (2013) Official statistics: Final UK greenhouse gas emissions, 5 February 2013. Department of Energy and Climate Change. https://www.gov.uk/government/publications/final-uk-emissions-estimates

DECC (2012) CCS Roadmap, Department of Energy and Climate Change, April 2012. [PDF] https://www.gov.uk/government/uploads/system/uploads/attachment_data/file/48317/4899-the-ccs-roadmap.pdf

Mackay, D. (2009) Sustainable energy without the hot air. UIT, 2 December 2008. http://www.withouthotair.com/download.html

Wiseman, J. and T. Edwards (2012) Post-Carbon Pathways: Reviewing Post-Carbon Economy Transition Strategies. University of Melbourne Sustainable Society Institute. CPD Occasional Paper 17.

3.10.3 Carbon omissions

Audsley, E. et al. (2009). How low can we go? An assessment of greenhouse gas emissions from the UK food system and the scope to reduce them by 2050. WWF-UK.

DECC (2013) Official statistics: Final UK greenhouse gas emissions, 5 February 2013. Department of Energy and Climate Change. https://www.gov.uk/government/publications/final-uk-emissions-estimates

DECC (2012) Alternative Approaches to Reporting UK Carbon Emissions. [Accessed 13 April 2013]. [PDF] https://www.gov.uk/government/uploads/system/uploads/attachment_data/file/79282/6745-alt-approaches-reporting-uk-ghg-emissions.pdf

Druckman, A., and Jackson, T. (2009) The carbon footprint of UK households 1990-2004: a socio-economically disaggregated, quasi-multi-regional input-output model. *Ecological Economics*, 68 (7), pp. 2066-2077.

Helm, D. et al. (2007) Too Good to be True? The UK's Climate Change Record. [Accessed April 15 3013]. http://www.dieterhelm.co.uk/node/656

Holding, J. et al. (2011) Food Statistics Pocketbook 2011. Food Statistics Branch: Department for Environment, Food and Rural Affairs, York, UK.

Wei, T. et al. (2012) 'Developed and developing world responsibilities for historical climate change and CO2 mitigation'. *Proc. Natl. Acad. Sci.*, 109 (32), pp. 12911-12915.

Chapter 4. Using ZCB

4.1 Changing how we think

Futerra (2005) Communicating Sustainability. United Nations Environment Programme.

4.2 Taking action in our homes, communities and places of work

Shepherd, A. et al. (2012) *The Home Energy Handbook*. Centre for Alternative Technology.

Chapter 5. ZCB and ...

5.2 ZCB and community energy

Devine-Wright, P. et al. (2007) An empirical study of public beliefs about community renewable energy projects in England and Wales. Working Paper 2: Community Energy Initiatives Project. [PDF] http://geography.lancs.ac.uk/cei/Downloads/PDW%20STP%20Working%20Paper%202.pdfS.

Harnmeijer, A. et al. (2012) A Report on Community Renewable Energy in Scotland. [PDF] http://scenetwork.co.uk/sites/default/files/SCENE_Connect_Report_Scotland.pdf

Rogers, J.C. et al. (2011) Social Impacts of Community Renewable Energy Projects: Findings from a Woodfuel Case Study, *Energy Policy* 42, pp. 239-247.

5.4 ZCB and health

DECC (2012) UK Greenhouse gas emissions, provisional figures and 2011 greenhouse gas emissions, final figures by fuel type and end-user. UK Government Department of Energy and Climate Change. [PDF] https://www.gov.uk/government/uploads/system/uploads/attachment_data/file/175659/ghg_national_statistics_release_2012_provisional.pdf

Hempel, S. (2006) *The Medical Detective: John Snow and the mystery of cholera*. Granta Books, London.

IMNA (2002) The future of the public's health in the 21st century. Institute of Medicine of the National Academies. National Academies Press.

Krieger, N. (2001) 'Theories for social epidemiology in the 21st century: an ecosocial perspective', *International Journal of Epidemiology*, 30.4, pp. 668-677.

Lang, T. and Rayner, G. (2012) Ecological Public Health: the 21st century's big idea? *BMJ*, 345:e5466, 21 August 2012.

QOF (2011) Diabetes in the UK – Key statistics 2012. Quality and Outcomes Framework (QOF) [PDF] http://www.diabetes.org.uk/Documents/Reports/Diabetes-in-the-UK-2012.pdf

Scarborough, P. et al. (2011) The economic burden of ill health due to diet, physical inactivity, smoking, alcohol and obesity in the UK: an update to 2006-07 NHS costs, *J Public Health (Oxf)*, Dec; 33 (4) pp. 527-35. DOI: 10.1093/pubmed/fdr033. [Live at: May 11 2011].

Wang, C. et al. (2011) 'Health and economic burden of the projected obesity trends in the US and the UK', *The Lancet*, Volume 378, Issue 9793, pp. 815-825, 27 August 2011.

5.6 ZCB and happiness

Hawken, P. (2007) *Blessed Unrest*. Viking Penguin.

5.8 ZCB and creative practice

British Museum (2013) Exhibited as part of Ice Age Art and the Making of The Modern Mind, Spring 2013.

Gablik, S. (2002) *Living The Magical Life: An Oracular Adventure*, p. 139. Phanes Press.

Heinzen, B. (2004) *Feeling for Stones*. http://www.barbaraheinzen.com/site/publications.php?catId=35&strCurrPath=B+-+Feeling+for+Stones&dirToBrowse=A+-+Excerpts+-+Feeling+for+Stones

Index

Adaptation: 7, 16, 121-122, 173, 179, 194
Adaptive capacity: 120, 124
Alternative Energy Strategy for the UK: 7
Amazon rainforest: 16
Ambient energy: 173
Anaerobic digestion (AD): 61, 66, 78-80, 97, 146, 172-173, 176
Arctic sea ice: 14-15, 180
Atmosphere: 13-14, 16, 18, 31, 33, 45, 47-48, 50, 52, 60, 68, 71, 74-75, 81, 85, 96, 102-103, 113, 121, 125, 133, 136, 157, 162, 173-175
Atmospheric aerosols: 18
Aviation: 25-26, 31, 36, 47-48, 50-52, 71, 72, 81, 95, 109, 114, 133-136, 184, 186
  domestic: 52
  fuel: 48, 50, 72, 133
  international: 25-26, 31, 36, 47, 52, 109, 114, 134-135

Back up generation: 173-74
Batteries: 50, 66, 70
Behaviour change: 159
Biochar: 32, 78-80, 98, 103, 105, 107, 122, 128, 173, 177, 191-192, 194
Biodiversity: 2, 16, 18, 31, 33, 36, 50, 83, 87, 97, 102, 105, 107, 109, 120, 122, 126, 133, 159, 173, 176-177, 191
  loss: 18, 126, 159
Biofuels: 50, 60, 71, 96, 126, 152, 173
  first generation: 71, 96, 126
  second generation: 71, 96
Biogas: 39, 45, 47, 54-55, 60-61, 66-70, 72, 74, 79-80, 95-98, 126, 173, 176-177, 186
Biomass: 32, 38-40, 43, 45-47, 50-56, 60-61, 63, 66-68, 70-72, 74-75, 78-81, 83, 85, 94-98, 102, 107, 126-127, 131-133, 146, 156, 173-174, 176-178, 187, 190
Buildings: 7, 9, 38-47, 68, 75-76, 80, 95, 97-99, 102-103, 107, 122, 127, 129, 136, 144, 176-177, 185, 192
Business as usual: 118, 166
Businesses: 8, 36, 73-75, 81, 98, 116, 164, 175, 187

Cap
  and Dividend (CAD): 116
  and share: 116, 173
  hard: 114, 116, 173, 175, 177
  soft: 114, 116, 174, 177
Car: 13, 48, 49, 52, 65, 69, 86, 141, 144, 152, 158, 163, 172
  share scheme: 141
  use: 86, 144, 152
Carbon
  allowances: 176
  budget (see below)
  capture and storage (see below)
  credits: 33, 113
  cycle: 100, 173, 174
  dioxide (CO_2): 13, 24, 31, 56, 66-67, 84, 171, 173, 175-177, 179, 181, 190
  emissions: 9, 31, 95, 103, 116, 119, 135, 140, 152, 159, 161, 174-175
  flow: 100-101, 173
  intensity: 173
  leakage: 115-116, 193
  management: 33, 109, 157
  neutral: 38-39, 41-42, 45, 50, 52-54, 60-61, 63, 67-72, 74-75, 96-97, 109, 127, 132, 173-174
  neutral synthetic gas (see Synthetic gas)
  neutral synthetic liquid fuel (see Synthetic fuel)
  omissions: 8, 31, 34, 42, 85-86, 90, 103, 115, 134-135, 196
  personal allowance: 116, 172
  price: 34, 114, 116
  Price Floor (CPF): 116
  store: 100, 102-103, 174, 177
  tax: 115-116, 119, 153, 174
  trading: 114-115, 119, 193
  zero: 8, 9, 30-32, 42, 49, 55-56, 68, 70, 81, 83, 95, 98, 106-107, 109, 112-113, 118, 130-131, 133-136, 138, 140-141, 143-144, 146, 151-156, 160-161, 165-167, 171
Carbon budget: 2, 25, 27, 30, 32, 34, 106, 111, 113-116, 121, 124-125, 133, 135-136, 147, 173, 175, 183, 193

global: 25, 27, 34, 111, 113-114, 121, 135
UK: 25-27
Carbon capture: 32, 102-103, 106, 109, 134, 172-173
and storage (CCS): 32, 75, 132
bio-energy carbon capture and storage (BECCS): 32
from air ('scrubbing'): 32, 134
Carbon emissions (see Carbon)
Centre for Alternative Technology (CAT): 8, 140, 167, 171, 195
Chemical pollution: 18
Climate change: 5-6, 8-10, 12, 14, 16-17, 19, 22, 24, 27-28, 30, 31-36, 42, 47-48, 50, 52, 81, 98, 100, 103, 106-108, 113-116, 120-122, 124-125, 127, 133-135, 139, 144-145, 152-154, 157, 159, 160-162, 164-165, 167, 172-173, 175-177, 179-189, 191-197
Committee on (CCC): 152, 185-186
dangerous: 12, 14, 19, 24, 113-114, 121, 179, 183, 193-194
extremely dangerous: 121
Levy (CCL): 117
mitigation (see Mitigation)
projection of: 26
Coal: 13, 36, 56, 74-76, 102, 143
Combined heat and power (CHP): 45, 60, 96, 174
Community: 2, 6-8, 10, 14, 16, 114, 129, 132, 138, 140, 147-48, 154-155, 157, 159, 163, 167, 174, 196-197
renewable energy: 154-155, 196, 197
Compost: 78-79, 104, 141, 174
Contrails: 48, 50, 52, 174
Cycling: 47, 89, 153

Dairy products: 36, 133, 176
Decarbonisation: 9, 10, 25-27, 32-34, 112-114, 129, 133-36, 141, 143, 147, 156-157
Demand management: 54, 65, 132, 174
Diet: 10, 82-83, 86-92, 94, 109, 129, 133, 159, 189-190, 197
Dispatchable generation: 66, 173-174
District heating: 60, 66
Draughtproofing: 42, 177
Drivers: 21, 152, 160
Drought: 13, 105, 122, 179, 180
Eatwell Plate: 89, 94
Economic
collapse: 12
costs: 19
growth: 19, 34, 124, 162
markets: 114
recovery: 118, 120
regeneration: 22
systems: 13
Ecosystem: 6, 18-19, 22, 118, 122, 126, 139, 148, 156, 173-174, 181
services: 156, 181
Education: 12, 128, 139, 144-146, 155, 159-162, 167
Electricity: 8, 38-40, 43-48, 50, 52-56, 58-61, 63-72, 75, 94-96, 109, 116, 131, 134, 140, 152, 154, 161, 172-177, 186, 199-200
generation capacity: 21
grid: 8, 43, 131, 174, 176-177
shortfall: 63-64
surplus: 54-55, 63-66, 68-72, 96, 176
Electrolysis: 45, 66-67, 70-71, 75, 173-174, 187
Emissions
agricultural: 84-86, 92
allowance: 116, 174
cap: 34, 174
consumption: 42, 174
global: 25, 33, 113
industrial: 175
non-energy: 41, 73-76, 81, 84, 98-99, 106, 108, 133, 136, 187
permits: 116, 173, 177
reduction scheme: 173-174, 176
reductions targets: 24-25, 112, 114-115, 146, 152
Trading Scheme (ETS): 116
trends: 14
UK: 31, 74, 81, 115, 134-135, 187
Employment: 8-9, 118-121, 126-128, 160-161, 195
Energy
conservation: 155
crops: 79, 94, 95-98, 103-104, 106, 190
demand: 21, 33-34, 37-49, 51-57, 59, 63, 65, 80, 94, 110, 122, 127, 136, 141, 174-175
density: 50, 70

efficiency: 44, 127, 154-155, 177
intensity: 44, 47, 175
mix: 54-55, 60-61, 122, 132
storage: 54, 63, 65-66, 70
supply: 10, 21, 33, 37, 54-57, 63, 65-66, 94-95, 110, 112, 118, 122, 131, 147, 174-175, 186
variable: 65
Energy use
cooking, lighting and appliances: 39, 41, 44, 46-47
cooling: 41, 43, 47
heating: 8, 39-47, 60, 63-66, 68, 105, 107, 116, 129, 133, 164, 174-177
hot water: 39-47, 60, 66
Environmental limits: 12, 22
Exports: 54, 68, 134

Farmers: 151-157, 164-165
Farming: 83, 156-157, 187
Fertiliser: 71, 74, 78-79, 84-85, 87, 89, 96-97, 99, 104-105, 126, 174-175
nitrogen: 85, 176
Fischer-Tropsch process: 175
Flood: 16, 102, 122, 156
Flooding: 122, 164
Flying: 31, 48, 51-52, 98, 114, 130, 133-134, 174
Food: 2-3, 7, 9, 13, 16, 18, 33-34, 36, 47, 50, 53, 60, 66, 71, 77-79, 81-90, 92-94, 96-97, 99, 105-106, 109, 122, 126, 133-135, 141, 144, 147, 156-157, 159, 162, 164-167, 171-173, 175, 178, 183, 185, 187-190, 192-196
crops: 71, 97, 122
high in fat, salt and sugar (HFSS): 86
high protein: 87-88
processing: 92, 157
production: 16, 36, 60, 71, 81, 83-87, 92, 133, 156-157, 166
starchy: 83, 90, 93
Forest: 36, 78, 83, 96, 98-100, 102-103, 105, 107, 109, 126-128, 177, 180, 184, 191-92, 195
harvested: 100
planting: 32, 102, 122
sustainably managed: 102-103, 177
unharvested: 102, 107
Fossil fuel: 12-13, 21-22, 32, 36, 39, 53-57, 60, 66-68, 70-71, 74, 84, 87, 96, 100, 102, 114, 116, 118-119, 129, 132, 134, 139, 146, 173-176, 178
Fracking: 21, 175, 182
Freshwater use: 18
Fruit and vegetables: 83, 90, 93
Fuel: 8-9, 13-14, 21-22, 32, 37-41, 45-53, 55, 60-61, 63, 66-72, 80-82, 84, 87, 90, 94-97, 99, 106-107, 109, 110, 116, 118-119, 122, 126-127, 129, 132-133, 136, 139, 146, 152-153, 162, 164, 167, 173-178, 183, 186, 190, 193, 197
mix: 37-39, 110, 175
poverty: 22, 118

Gas
natural: 36, 39, 45, 66-67, 70, 76, 132, 140, 164, 176
synthetic: (see Synthetic gas)
Genetically modified (GM) crops: 87, 172, 188
Geoengineering: 32-33, 113, 125, 134, 136, 183-184, 195
Geothermal
electricity: 60-61
heat: 43, 56, 60-61, 87, 177
Global average temperature: 14, 17, 25-26, 30, 111, 113, 136, 175
Grassland
intensively grazed: 18, 94, 97, 176
semi-natural: 85, 87, 94, 97, 103, 105, 177
temporary: 107, 177
Greenhouse effect: 75-76, 173-176
Greenhouse gas emissions: 13, 35-36, 84, 86, 109, 174, 176-177, 183, 184-189, 191, 196-197, (see also Emissions)
Greenhouse gas (GHG) (super): 74, 176-177
Green New Deal: 118-119, 194
Gross domestic product (GDP): 19, 23, 172

Happiness: 21-23, 162-163, 182, 197
Happy Planet Index (HPI): 129, 172, 182, 195
Health: 2, 9, 12, 16, 18, 22-23, 49, 83-84, 86-87, 89, 92, 94, 128, 158-159, 162, 167, 188-190, 197
ecological public: 158, 197
temporary: 107, 177
Heat: 13, 16, 39-43, 45-47, 54-56, 60-61, 63, 65-66, 69, 74, 76, 87, 95-98, 122, 125, 131,

172-177, 180-181, 185, 199-200
loss: 41-43, 176
pump: 39, 43, 47, 55-56, 60-61, 63, 66, 76, 131, 173, 175
recovery ventilation: 42, 175, 177
store: 45-47, 66, 175
stress: 122, 175
wave: 16
Historical responsibility: 10, 26, 113-114, 139, 175
Household: 8, 22-23, 34, 36, 43, 58, 60, 65-66, 73-75, 78, 81, 86, 98, 108, 116, 133, 155, 172, 177, 184, 187, 194, 196, 201
low-income: 116
occupancy: 34
Hydropower: 7, 59-61, 66, 122, 131, 154, 175

Imports: 21, 33-34, 68, 83, 86-87, 93, 118, 126, 132, 134-136, 174, 177
Industrial: 2, 6, 13-14, 18, 24, 26, 32, 34, 36, 39-42, 44-47, 58, 60, 65, 70, 73-75, 81, 95-96, 103, 108, 113, 126, 133, 136, 175-177, 186-187, 191, 199-200
output: 41-42, 44, 47, 74-75, 176
processes: 36, 41, 44, 47, 65, 70, 73-74, 95, 103, 108, 133, 175
Industry
cement: 41, 74, 76
heavy: 126
iron and steel: 41, 44, 74-76
Inequality: 12, 23, 162, 182
Infrastructure: 13, 21, 28, 39-40, 49-50, 52, 63, 65-67, 70, 76, 96, 98, 102-103, 107, 112, 115, 119, 122, 136, 157, 159, 173, 176
Insulation: 42, 47, 103, 112, 122, 155, 176-177
International credits: 33, 113, 134, 136
IPCC - Intergovernmental Panel on Climate Change: 9, 13-14, 32, 100, 105, 172, 179, 183, 191
Irrigation: 164

Jobs: 9, 44, 86, 118, 121, 126-128, 136, 160-161
Justice: 6, 158, 163-165

Kyoto Protocol: 24-26, 114, 134-135, 176, 182

Land: 2-3, 7, 9, 13, 18, 32-33, 35-37, 38, 50, 52, 57-58, 60, 63, 71-72, 75, 79, 81-90, 92-99, 103, 105-110, 112, 115, 122, 124-127, 133-136, 144, 147, 156-157, 164, 171-173, 175-176, 178-179, 184-185, 188, 190-192, 195
agricultural: 18, 36, 77-79, 81-88, 90, 92, 97, 99, 103-107, 109, 122, 127, 157, 164, 180, 184, 188, 189, 190, 192, 195
management: 81-82, 99, 103, 112, 192
use change: 71, 85-87, 90, 98-99, 106, 126, 134-135, 156
use change abroad: 85, 134-135
Landfill: 73, 76-77, 80, 187, 190
Lifestyle change: 19, 28, 86, 120, 129, 132-133, 140, 156, 161, 163, 166
Livestock: 33, 36, 78-79, 82-85, 87-88, 90, 92, 94-97, 105, 107, 109, 125-127, 133, 135, 156-157, 176-177, 189, 191, 194-195
grazing: 82-83, 85, 87-88, 92, 96-97, 105, 109, 125-126, 133, 188

Meat products (lab produced): 89
Methane: 13, 16, 31, 45, 66-68, 70, 72, 74-77, 79-80, 84-85, 87-89, 92, 171, 173, 175-177, 187, 188-190
Miscanthus: 61, 95-97, 110, 176, 190
Mitigation: 19, 121, 135, 156, 176, 183, 188-189, 191-192, 196

new economics foundation (nef): 129, 181-182, 194, 196
Nitrogen: 18, 85, 89, 92, 97, 126, 172, 174, 175, 176, 189-195
flow: 18
inhibitors: 89, 172, 176
Nitrous oxide: 13, 31, 74, 84-85, 171, 174, 176, 189
Nuclear: 7, 32, 58, 63, 66, 116, 122, 132, 146, 194
Nutrients: 71, 78, 86, 97, 105, 122, 175-176, 188
Nutrition: 6, 7, 92, 172, 188

Obesity: 23, 83, 94, 158, 189, 190, 197
Ocean: 13-14, 16, 18, 121, 125-126, 173, 176, 179, 195
acidification: 16, 18, 33, 120, 125, 176, 195

Oil
conventional: 176
peak: 116, 120, 139, 176, 193
unconventional: 102-103, 175, 178
Ozone depletion: 18-19, 125

Passivhaus: 42, 46, 176
Peatland: 36, 86, 98, 104-105, 107, 122, 133, 176, 184, 190-191, 194
Permafrost: 16, 176, 180
Personal Carbon Allowances: (see Carbon)
Petroleum: 48, 50
Phosphorus flow: 18
Planetary boundaries: 17, 19, 120, 124-126, 181, 194
Policy: 2-3, 8-9, 25-26, 33-34, 78, 111-116, 140-141, 143, 153, 157, 159, 164-165, 167, 174, 179, 182-183, 186-189, 191-193, 197, 202
downstream system: 114
framework: 114-116, 174
global: 25
international: 3, 113, 140
local: 114
mechanism: 115-116
scheme: 114, 118
UK: 112-114
upstream system: 114
Population: 3, 25-26, 33-34, 39-40, 42, 46, 47, 82-83, 87, 90, 92, 94, 124, 133, 148, 152, 158, 164-165, 173, 176, 181, 184, 195
growth: 47, 124
Poverty: 22, 118, 158, 163-165
Protein: 82, 86-90, 92, 97, 126, 189
Protein (animal-based): 87
Protein (plant-based): 87, 90
Pumped storage: 66
Pyrolysis: 78, 105, 173, 177, 191

Rail: 13, 47, 49, 50-53, 122
Recycling: 44, 47, 74, 76-80, 161-162, 187-188
Renewable energy: 7-8, 10, 32-33, 38, 42, 44, 46-47, 54-57, 58, 60, 63, 65-66, 68, 70, 94, 108, 118-119, 122, 126-127, 154-155, 161, 171, 186, 193, 196
Renewables: 38, 45, 50, 55-56, 63, 65-68, 118, 126-127, 131-132, 147, 154-155, 177, 186
Resilience: 31, 129, 154
Retrofit: 42, 177

Sabatier process: 67, 96, 174, 177
Sea level rise: 17, 122, 179
Shipping: 25-26, 31, 36, 47-48, 51, 109, 114, 134, 135-136
freight: 53
international: 48
Short Rotation Coppice (SRC): 61, 95-97, 177
Short Rotation Forestry (SRF): 61, 83, 95-97, 107, 177
Smart
appliances: 45, 47, 69, 177
car charging: 65, 69
Soil: 83, 85, 87, 89, 92, 97, 102-106, 174-177, 189, 191-194
carbon: 103-105, 177, 192, 193
Solar: 6-7, 13, 39-40, 43, 47, 56, 60-61, 65-66, 68, 126-127, 131-132, 155, 161, 173, 177
photovoltaics (PV): 60
thermal: 39-40, 47, 56, 60-61, 66, 177
Sustainability: 17, 87, 126, 144, 148, 158-161, 163, 171, 177, 192, 196
Sustainable development: 19, 164
Synthetic
fuel: 52, 61, 71-72, 176, 177
gas: 39, 45, 47, 54-55, 61, 63, 66-71, 74, 76, 94-98, 108, 127, 174

Tidal: 56, 59, 61, 131, 144, 177
range energy: 59, 177
stream energy: 59, 177
Tradable Energy Quotas (TEQs): 116, 177
Transport: 2, 7-9, 36, 38-39, 45, 47-53, 55-56, 60, 63, 66, 68, 70, 72, 81, 84, 89, 94-96, 109, 115, 122, 126, 127, 133, 135, 141, 143-144, 147, 152-153, 158-159, 166, 184-186
public: 47, 49, 51, 141, 152-153
sustainable: 152-153
system: 47, 49, 53, 81, 135, 152

UK Climate Impacts Programme: 121, 172
United Nations Framework Convention on Climate Change (UNFCCC): 19, 24, 31, 114, 182

Variability: 63, 147, 179, 180, 194
Vehicles: 47-53, 65-66, 69-72, 84, 95, 133, 152-153
  electric: 47, 50, 52, 65-66, 152-153
  heavy commercial: 50, 52-53, 70-72, 84, 95, 133
  hydrogen: 47, 52

Walking: 47, 89, 153
Waste
  agricultural: 78-79
  biodegradable: 80
  food: 66, 77-79, 92, 141
  incineration: 76
  management: 7, 73, 76-77, 81, 98, 108, 161
  water: 77, 79
Wave power: 59, 61
Welfare: 162, 178
Wellbeing: 8, 9, 21-23, 30, 84, 120-121, 128-129, 162, 177-178, 182, 195
Wind: 6-7, 13, 38, 54-61, 63, 65-66, 68, 70-71, 76, 112, 126-127, 131-132, 144, 146-147, 154, 175-177, 186
  offshore: 54, 56, 58, 60-61, 71, 112, 126, 147, 175-176
  onshore: 58, 61, 132, 176
  speed: 57-58, 63, 65-66, 68, 144
Wind turbine
  fixed offshore: 58, 175
  floating offshore: 175-176
Wood products: 47, 79, 83, 99, 100, 102-103, 105, 107, 127-128, 136, 177
World Health Organisation (WHO): 190
World Trade Organisation (WTO): 115

ZeroCarbonBritain 2030: A new energy strategy: 195
zerocarbonbritain: An Alternative Energy Strategy: 8

Zombies: 148